Advances in Military Geosciences

Military activities are almost always strongly integrated within a wide spectrum of geoscience. The decisive outcomes of land battles throughout history have been dictated in large part by the terrain and environmental setting. Modern military operations rely on a wide range of land-, air-, sea-, and space-borne intelligence and knowledge of dynamic terrain processes and conditions. In addition, the study of geo-based environmental science is critical to both the sustainable management of military reservations and installations, as well as the evaluation of how terrain and environmental conditions may impact military equipment and operations.

Advances in Military Geosciences contains single and multi-authored books as well as edited volumes. Series Editors are currently accepting proposals, forms for which can be obtained from the publisher, Ron Doering (ron.doering@springer.com).

More information about this series at http://www.springer.com/series/15030

Stuart Pearson · Jane L. Holloway
Richard Thackway
Editors

Australian Contributions to Strategic and Military Geography

Springer

Editors
Stuart Pearson
School of Physical and Environmental Science
University of New South Wales
Australian Defence Force Academy
Canberra, ACT, Australia

Jane L. Holloway
Joint and Operations Division
Defence Science and Technology Group
Department of Defence
Canberra, ACT, Australia

Richard Thackway
School of Physical and Environmental Science
University of New South Wales
Australian Defence Force Academy
Canberra, ACT, Australia

ISSN 2522-8315 ISSN 2522-8323 (electronic)
Advances in Military Geosciences
ISBN 978-3-030-10372-9 ISBN 978-3-319-73408-8 (eBook)
https://doi.org/10.1007/978-3-319-73408-8

Softcover re-print of the Hardcover 1st edition 2018

Printed on acid-free paper

This Springer imprint is published by the registered company Springer International Publishing AG part of Springer Nature
The registered company address is: Gewerbestrasse 11, 6330 Cham, Switzerland

Contents

Chapter 1
Introduction

Stuart Pearson, Jane L. Holloway, and Richard Thackway

There is a moment, in the emergence of a new ontological or epistemological approach, when a potential transformation is framed by the past, informed by the present and inspired by potential futures. Australian strategic and military geography may be at that moment. This book presents the case for a resurgence of strategic and military geography in Australia, principally by examining a few conceptual aspects, a first approximation of a history, indicative contributions from some specific domains of expert knowledge and some recommendations for future developments.

Australian contributions to strategic and military geography are diverse and eclectic; encompassing matters of interest in the strategic, operational and tactical domains; from micro to macro; theory and practice. Some inform and contradict each other. These are essential parts of an integrative role and the contest of systemic coherence in the geographical discipline.

In the writers' collective opinion, Australia has contributed, and will continue to contribute, to strategic and military geography. Geography remains fundamental in today's world – particularly in a defence and security context – and even at transmission speeds faster than humans can engage with. The strategic, operational and tactical worlds change relative to expectations people have about defence and security.

This contribution was inspired to present, and nudge, different ways of thinking about the changes in Australia's defence and security environments. The context of

S. Pearson (✉) · R. Thackway
School of Physical and Environmental Science, University of New South Wales, Australian Defence Force Academy, PO Box 7916, Canberra 2610, ACT, Australia
e-mail: s.pearson@unsw.edu.au

J. L. Holloway
Joint and Operations Division, Defence Science and Technology Group, Department of Defence, Canberra, BC 2610, ACT, Australia

S. Pearson et al. (eds.), *Australian Contributions to Strategic and Military Geography*, Advances in Military Geosciences,
https://doi.org/10.1007/978-3-319-73408-8_1

this volume about Australia's contribution to strategic and military geography is strongly shaped by the people and discussions that brought it together. These contributions include the emergence of 'wicked problems' and the increasing use of a systemic or holocentric heuristic in response, the rise of cyber, the increasingly intrusive connectedness of human endeavours around the globe and emerging Earth system reactions, leading to non-linear changes in biophysical, human and cyber geography (Holloway 2017, Chap. 4). These are accompanied by increasing complexity, an emerging multipolar world, increasing capabilities in computing and communication, force asymmetry, stand-off methods of war, confidence in design thinking, artificial intelligence and paradoxes in security, connectedness, prosperity and health. An often neglected, but nevertheless vital part of our strategic and military geography is the degree to which acquisitions of land-based assets and periodic ecologically-based assessments of their extent and condition are meeting the current strategic and projected future training needs (Thackway and Pearson 2017, Chap. 15).

Each chapter is a standalone contribution and collectively the chapters share numerous and common themes, memes and concepts in terms of strategic and military geography. Viewed as a whole, the chapters are variegated and reflect differences in formality, style and accessible detail. This volume as a whole presents observations and ideas representing a wide scope and diversity of skills and expertise across numerous disciplines. It characterises current thinking in Australian strategic and military geography.

The authors comprise researchers from universities, industry, former and serving military officers and public (or civil) service policy makers. Their initial purpose was to share and discover what was of interest and prospective for further discussion and research. Australia, compared to the US and UK, has a small inconspicuous strategic and military geography sub-discipline at this time. This community were keen be informed of what is happening elsewhere and to become part of a larger conversation and community.

This book describes, without being excessively narrow or exceptionalist, Australian contributions to strategic and military geography. Our contention is that Australia has further contributions to make. We seek to demonstrate it in this book, in the material the authors cite, in their observations and what they recommend. The book provides applied knowledge, and each author contributes points that are actionable and written for readers with military, public or academic perspectives.

1 The Australian Way

Australia has a highly variable and dynamic environment making it difficult to generalize about its geography or its biophysical or human characteristics (Thackway and Ford 2017, Chap. 11). Australia's land and sea estate spans the tropics to the South Pole, 5% of the world land and 10% of the world ocean area. Australia has a multicultural society spread across many regional bio-climates and human-environment sub-systems. It has global strategic interests and it also has an interest in the increasingly obvious global, regional and continental impacts of collective human activity. Globalisation and other connectedness is changing the world: overwhelming some Earth system boundaries; driving cybernetic and technological singularity with upside and downside risks; and challenging our old ways of thinking, finding out, taking action and evaluating the results may be part of the problem rather than the solution (Holloway 2017, Chap. 4). The drivers for new ontological and epistemological insights raised by authors in this volume concern strategic and military matters, but they are also set within a larger (global and transformational) context and also in a small, very human, set of constraints.

Continental Australia has supported indigenous peoples for over 60,000 years. Those original inhabitants, who have the oldest living cultures in the world, shaped the continent's social and biophysical landscape into the biggest estate on earth (Gammage 2011). The perspectives of these original peoples and their ancestors are not routinely incorporated into modern Australian society. Knowledge of their geography and history has been variously misrepresented, disregarded and suppressed. Their cultures, languages and knowledge, which are overwhelmingly oral and place-based and closely linked into the biophysical landscape, have often been misunderstood or deliberately misappropriated since the advent of European settlement (Henry Reynolds 2017, Chap. 12). With the overturning of the 'terra nullius' (literally empty land) precept in Australian law, Aboriginal and Torres Strait Islander peoples are increasingly able to reconnect to their ancestral lands. Conversations about what are described as Frontier Wars – continued in this volume – are becoming better understood and accepted. That is not to imply that the Frontier Wars are over. Australian institutions and legal arrangements, mostly derived from the United Kingdom (mainly England), still retain inherent biases against the ancient cultures that pre-dated Britain's forcible acquisition of the continent. A lesser violence continues in Australia that results in the starkly inequitable inter-generational outcomes for Australia's indigenous peoples, much of it related to social and economic marginalisation, loss of language, culture, place and space, and disconnection from their ancient songlines and dreaming ancestral dreamtime and country.

When the British Empire established their port cities in Australia from 1788 they did so under their flag, used the para-military forces of the state and the immigrants to suppress the indigenous people and the environment to make their systems of land tenure, race, religion and agriculture work. Australian environments (of course including the social environment) were reshaped in novel ways – imposing private land ownership, building towns and cities, developing the land by clearing trees, impounding waters and establishing intensive agriculture.

New arrivals, including some outcast criminals and some trained in the military who had served in foreign wars, followed a colonial path and contributed to the culture of frontiersmen, bushmen and militarists in Australia. Distance from the Empire was in many ways the defining issue of Australian human, cultural and physical geography (Blainey 1966). Australia started a to become more independent in thought and action with remarkable changes shaped by new geographies. The events and processes of developing Australia's Anglo-European human geography are already well described.

Today Australia's prosperity – geographically, economically and socially - is best served by an open, globally-enabled trade economy with air-sea borders that suit a wealthy and multicultural population. Australians pride (and sometimes delude) themselves on their individualism, anti-authoritarian larrikinism, egalitarian mateship, tolerance of cultural difference, commitment to a fair go, freedoms of speech and privacy, and human rights. At scales beyond the nation state, Australia strongly supports the UN system and was a significant actor and supporter in its creation (Pearson 2017, Chap. 2 and Press 2017, Chap. 9). Australia has a self-narrative of being a good international citizen, a middle power that "punches above its weight", and an honest broker who, because of its limited military power and geographic position, contributes a useful voice to the rules-based global order.

Australia's strategic and military position has continued to change. The dominant mantra is about being a middle power, a good neighbor and participant in the rules-based global order with a focus on its geographical region. Australia's interests are dictated by its: English-speaking legal, political and other cultural heritage; position as a globally competitive supplier of commodities (coal, iron, education, food and fibre) – especially to China and Japan; location in the Asian sphere; maritime isolation and responsibilities; reliance on imported products and fuels; and enduring alliances with the USA, United Kingdom, New Zealand and Canada, especially in the areas of Defence and security. Australia also maintains close military and other ties with neighbours such as Indonesia, Papua New Guinea and Singapore particularly.

2 A Future of Contribution

Australians use mobs of idioms – one, "she'll be right" - is used to describe a (possibly naïve but droll) optimism, apathy or an acceptance of what might be called 'a fit for purpose' standard (rather than perfection). The use of the feminine pronoun in that idiom might further suggest that the future may be fickle or productive; unpredictable and lucky. Australians have changed: The old bush-city banter of Lawson and Paterson in the pages of national papers (1892–3) is nothing more than an artifact to most contemporary urban Australians. Beyond immediate interest in North Korea's nuclear and intercontinental missiles, the scourge of home-grown extremism and terrorism in the region - Australia's near-term future view includes: China's and India's emergence as military powers with strategic needs and maritime ambitions; African nations' ongoing struggles; failing states in the region, megacities and insurgent wars; and the emergence of climate change-strengthened natural disasters that appear to be on a pathway to become an existential threat (DoD 2016). Future strategy and military work – including Australian Defence Force (ADF) deployments – will require additional technical and cultural literacy. It is our contention that geography's methods, skills and attitudes are more useful than they have been for nearly 60 years when the world was at war and geographers were mapping the islands of Australia's northern approaches.

3 Australia's Defence and Security Environment – New Frames

Two important documented processes frame the ADF; numerous Defence reviews and the Defence budget; both are frequently published and widely accessible. Yet reading the policy or the doctrine is often less informative than following the money (Thomson 2017). Australia's expenditure on Defence is over 34 billion dollars or 1.9% of Australia's GDP and that is, according to a recurrent report, Ninety-five million, thirty-two thousand, five hundred & ninety-one dollars & seventy-eight cents per day (Thomson 2017). Rapid growth in preparedness required and new capabilities strains the organization's ability to spend the money. New procurements of submarines, fast jets, cyber security and other domestic security measures are growing with proposals for stronger homeland security arrangements.

Defence is review rich – it has had more than 36 reviews since 1973 (Conroy 2015), one of the most recent is *The First Principles Review* that recommended it become a 'holistic, fully integrated One Defence system' centralized to strengthen accountability and top-level decision-making (DoD 2014). Judgement on its failings will come from an Audit of its performance due in March 2018 (ANAO 2018). Failures, inefficiency and poor accountability characterise previous reports. A relatively large civil-military organisation is a cumbersome thing to change in peace time and perhaps even harder when there is a high tempo of overseas deployments (Pearson 2017, Chap. 2).

The development and implementation of the Australian Defence White Paper 2016 involved some fundamental rethinking about Australia's geographical place in the modern context of a rising China, climate change and cyber security. Defence White Papers are Australia's principal public policy document regarding defence and security. They present the Government's assessment of Australia's strategic environment, long-term strategic direction and commitments for Defence (Table 1.1), as well as setting out future capability requirements. White Papers include policy guidance on strategy, capability, industry, innovation, posture and international engagement.

The detailed strategic analysis of experts such as Alan Dupont (Dupont and Reckmeyer 2012; Dupont 2015), Paul Dibb (Dibb 2006), Hugh White (White 2012), Adam Lockyer (Lockyer 2017) and others cannot be synthesized here – in summary there is a smorgasbord of alternatives and a shortage of diners prepared to commit themselves. Choice-making and then investing adequately and staying the chosen course appear as central criticisms of Defence and of the Australian political system more generally.

Table 1.1 Australia's strategic defence framework (DoD 2016)

Strategic Defence Interests		
A secure, resilient Australia, with secure northern approaches and proximate sea lines of communication.	A secure nearer region, encompassing maritime South East Asia and the South Pacific.	A stable Indo-Pacific region and a rules-based global order.
Strategic defence objectives		
Deter, deny and defeat attacks on or threats to Australia and its national interests, and northern approaches.	Make effective military contributions to support the security of maritime South East Asia and support the governments of Papua New Guinea, Timor-Leste and of Pacific Island countries to build and strengthen their security.	Contribute military capabilities to coalition operations that support Australia's interests in a rules-based global order.

Australia's current contribution to the broader global Defence and the security environment can be understood using three contemporary frames. These emerging requirements are not comprehensively covered in this volume but we hope they will be in future applied to military geography research.

3.1 Frame 1 – Cyber Geography and the Security Boomerang

Cyber Geography has emerged over the past 15–20 years (e.g. Kitchin 1998) and it is being increasingly securitised, including through the development of Australia's first Cyber Security Strategy (Department of Prime Minister and Cabinet 2016). In 2016–7, following well-reported global cyber security incidents, the Australian Signals Directorate, situated in Defence, was directed "to use its offensive cyber capabilities to disrupt, degrade, deny and deter organized offshore cyber criminals" (Payne 2017). Use of this capability was previously limited to helping target, disrupt and defeat terrorist organisations. This new directive maintains consistency with Australia's international legal obligations, and is subject to stringent legal oversight. And yet it also illustrates the 'boomerang effect' of capabilities being deployed abroad by the military subsequently 'coming home' and crossing into civilian life even when they may be at odds with Australian values (Kilcullen 2016).

3.2 Frame 2 – Homegrown Threats

Endogenous threats; home grown militants, returning nationals who have fought in foreign conflicts, civil emergency, natural disasters and humanitarian crises have changed the expectations for the kinds of tasks the military are allocated. The increasing likelihood of extreme and unrelenting climate and other Anthropogenic changes to Earth systems, the faith in the military to work when all else fails (an idea described clearly for the US by Brooks 2017) and an awareness (often beyond the capability to respond) that complex human and biophysical systems are at work.

In addition, there is an emerging global and scientific consensus that planetary boundaries, that have provided a relatively safe and stable operating space for the human race for thousands of years, are being progressively and irreversibly exceeded. These complex, intertwined and novel challenges have obvious – and cascading - consequences to defence and security. The implications at local, national, regional and global scales involve the United Nations, member governments and sub-national groupings (cities, local or provincial governments) who are increasingly mobilizing to address national responses. Sometimes these include whole-of-government activities and military organisations, some of which – like the US Navy – have led identification and analysis of impacts. The ADF is designed and indoctrinated to fight foreign (and mainly state-based) threats – is it appropriately equipped and tasked to extend the concepts of defence and security to potentially catastrophic changes in (global and regional) biophysical geography?

3.3 Frame 3 – Integration and Applied Science

Geography, as an integrative and applied science, is able to use the tools and understanding of human and biophysical sciences. It draws on the knowledge of many disciplines, to solve and resolve problems that defy discipline-specific solutions. Working with complex adaptive systems and system-of-systems approaches provides an engaging, effective and appropriate way to consider global change (Holloway 2014). Strategic and military geography offers *tools* to know place and space, *time* and *scale* that can be used to identify and work with systems of interest, and purposeful application to achieve desirable outcomes. While not specifically referencing the discipline of strategic and military geography, the ADF uses equivalent tools, systems, disciplinary experts and teams to achieve end states through applying their own conceptual understandings. The ADF's involvement, along with its allies, in recent decades with disaster relief, humanitarian assistance, peacekeeping and counter-insurgency warfare (Kilcullen 2017, Chap. 7) have honed these skills, doctrines and resources.

Geography has a strong focus on places and it also has highly permeable boundaries – it facilitates innovation, cross-disciplinary work and often seeks application and feedback from practitioners. These characteristics are demonstrated in the chapters of this book. The use of geography in war is ancient and it forms one of the foundations for all military education and training. Much of military geography is highly transferable through history and internationally. Australian targeting doctrine, for example, includes quotations by Sun Tzu, a strategist from 2500 years ago and 200-year-old advice from Clausewitz. Some aspects of military geography, however, are born of perspectives closely shaped by a country's geography, its resources and people, and its modern – and ancient – history.

4 Reviving Geographical Concepts

Landscapes and geography are foundations of human experience, learning and discussion in society. Geography is compulsory in Australian primary and secondary education (Holmes 2015) and everyone is expected to be a geographical thinker. Others, sometimes former geography graduates but also from agricultural and forestry systems disciplines, go on to apply the language and insights of geography in strategic studies, humanities or engineering. People use geography's essential understandings of spatial arrangement as a lens and use the traditional strengths of inter-disciplinarity, scalar analysis, synthetic capabilities, acceptance of multiple knowledges (including traditional, scientific, religious, modelled) and a willingness to be (self-) critical and constructive.

Holmes (2002, 2016) and Pearson et al. (2005) have described the challenges of being geographers in Australian universities. In summary, coupled human and biophysical academic traditions in geography in Australia delaminated during the 1960–80s, with social and cultural geographers making their connections to the humanities, critical or Marxist approaches, and physical geographers finding stronger affinities to physical sciences and environmental systems disciplines.

In Australian universities, former geography disciplines have split into humanities and physical sciences and now these disciples are rarely in the same organizational unit. Integrated human and biophysical analyses are now typically being done as isolated units without reference to 'Geography' for reasons suggested by Holmes (2002). The proposal explored by the authors in this book is part of a nascent movement to re-engage across disciplines and agencies to share geographically-related insights and support people working on places, problems, science and arts of military and strategic geography.

A brief overview of the history of Australian contributions to the sub-discipline of military geography provides some initial structures to consider (Pearson 2017, Chap. 2). An historical approach might also help ensure only new mistakes are made (rather than repeat the same ones) and help build on previous thinking rather than restarting the work. Australian Geography's academic foundation is largely attributed to Griffith Taylor who wrote, often controversially, about Australia's population, especially its distribution, and other human and biophysical geography issues.

Strategic Military Geography has developed as a heuristic model for discussing and preparing for military responses to global changes; including biophysical, climate, biodiversity and land use since 2012 (Holloway et al. 2015). That model was highly successful through engagement of senior Defence force planners and policy makers. It continues to be revised and improved (Holloway 2017, Chap. 4) for use by Defence and with other military organisations, other government agencies, researchers and industry. A particular focus in this endeavour is the likely consequences for defence preparedness (readiness) and future operations. Military preparedness requires response options to global biophysical, socio-cultural and cyber-geographic changes, including those provoked by climate change, biodiversity loss, land use change, aerosols and synthetic chemicals.

5 'Looking out' and 'Looking in' – Perspectives of the Authors

The contributions presented in this book provide two broad perspectives on strategic and military geography. Some authors have taken the vantage point of 'looking out' from Australia's shores (e.g. Peter Kenshole, Greg Austin, David Kilcullen), while other authors have taken a perspective of 'looking in' on Australia (Michael Thomas, Tony Press, Henry Reynolds and Stuart Pearson, Richard Thackway and Fred Ford). For example, (Richard Thackway and Stuart Pearson 2017, Chap. 15) explore the application of well-developed civilian technology and assessment frameworks into the Defence estate to meet the increasing demand for Defence to be an environmental Estate manager and to prove that publicly.

Other chapters are agnostic to these two perspectives and take global perspectives (e.g. geospatial intelligence by Rob Coorey (2017, Chap. 10); hyper-threats by Elizabeth Boulton (2017, Chap. 5); and security implications of El Nino by Michael Thomas). Coorey's experience as a Defence-related business is brought to life in his work on GEOINT, a domain of rapid technological advance and civilian application. In a more philosophical way, Boulton's chapter (2017, Chap. 5) has developed new ways to conceptualise the challenges of climate change. Boulton's hyper-threats show the power of a team and book authoring project to drive discussions between disciplines. Thomas's chapter (2017, Chap. 13) explores global circulation that creates security teleconnections that can worsen or provide opportunities in the ADF's strategic and operational environments. This kind of work is very useful to an agency that builds capabilities years ahead. Together Oppermann and Brearley (2017, Chap. 14), Boulton, and Thomas provide inputs to the broader consideration of climate change and ground their research in their military practice, thus providing some original Australian contributions to the resurgent sub-discipline of strategic and military geography.

Dan Cassidy's (2017, Chap. 6) examination of cultural and spiritual differences, between Australian and our near neighbours, provides strategic insights for ADF deployments. Tony Press (2017, Chap. 9) shows the strategic nature of Australia's scientific work in Antarctica and suggests this collaborative approach to national interaction can deliver Australia's interests. While Henry Reynolds (2017, Chap. 12) shows the value of synthesizing his knowledge of Indigenous history and the Frontier Wars providing valuable insights into how Australia's cultural and physical geography influenced the range and nature of those wars.

6 Where to from Here?

The sub-discipline of strategic and military geography is loosely drawn together in Australia using an informal study group, associated with the Institute of Australian Geographers. There are connections to global communities such as the American Association of Geographers, the British Association of Geographers and the European Geophysical Union. This will enable the conversations started in and by this book to be continued and expanded into the future. Some specific considerations for a strategic and military geography community, both locally and internationally, include:

1. Definitions and key tenets of the sub-discipline and relationships to other disciplines and approaches including regional geography, systems science and learning philosophies.
2. Leadership – intellectual and practical – and community building.
3. Community engagement to elicit key questions for future research.
4. International collaborations

7 Conclusions

The re-emergence of military geography, particularly at the strategic level, in Australia is described in this book using conceptual, historical and empirical evidence. It shows that Australia has contributed expert knowledge, skills and an attitude that promises future development in geography. The authors share a belief in the potential for geography to provide boundary-crossing opportunities for future contributions and have helped identify the domains that look particularly prospective.

References

ANAO. (2018). *Defence's implementation of the First Principles Review*. Canberra: Australian National Audit Office.

Blainey, G. (1966). *The tyranny of distance: How distance shaped Australia's history*. Melbourne: Macmillan.

Brooks, R. (2017). *How everything became war and the military became everything: Tales from the Pentagon*. New York: Simon and Schuster.

Conroy, P. (2015). *The First Principles Review—One Defence, but at what cost? The Strategist blog*. Canberra: Australian Strategic Policy Institute.

Department of Prime Minister and Cabinet. (2016). *Australia's cyber security strategy*. Canberra: Department of Prime Minister and Cabinet.

Dibb, P. (2006). Is strategic geography relevant to Australia's current defence policy? *Australian Journal of International Affairs, 60*(2), 247–264.

DoD. (2014). *First principles review – Creating One Defence*. Canberra: Department of Defence.

DoD. (2016). *Defence White Paper*. Canberra: Department of Defence.

Dupont, A. (2015). The next defence white paper: Challenges facing defence. *United Service, 66*(1), 14–16.

Dupont, A., & Reckmeyer, W. J. (2012). Australia's national security priorities: Addressing strategic risk in a globalised world. *Australian Journal of International Affairs, 66*(1), 34–51.

Gammage, W. (2011). *The biggest estate on earth: How Aborigines made Australia*. Sydney: Allen & Unwin.

Holloway, J. (2014) *Assessing global change impacts on Defence using a System-of-Systems perspective*. Proceedings of the 9th IEEE System of Systems Engineering (SOSE) Conference, Adelaide, Australia. pp 260–265 doi https://doi.org/10.1109/SYSOSE.2014.6892498.

Holloway, J., Thomas, M. D., & Durrant, C. (2015). *Strategic military geography: Climate change adaptation and the military* (pp. 493–514). Handbook of Climate Change Adaptation: Springer.

Holmes, J. (2002). Geography's emerging cross-disciplinary links: Process, causes, outcomes and challenges. *Geographical Research, 40*(1), 2–20.

Holmes, J. (2016). Whither geography? A response to Finlayson's concerns. *Geographical Research, 54*(1), 103–106.

Kilcullen, D. (2016). *Blood year: Islamic State and the failures of the war on terror*. Melbourne: Black Inc.

Kitchin, R. M. (1998). Towards geographies of cyberspace. *Progress in Human Geography, 22*(3), 385. https://doi.org/10.1191/030913298668331585 1998 22: 385.

Lockyer, A. (2017). *Australia's defence strategy: Evaluating alternatives for a contested Asia*. Melbourne Univ Publishing Melbourne.

Payne, M. (2017). Capability to fight cyber criminals. Media release Friday, 30 June 2017. Accessed 12 Aug 2017 http://parlinfo.aph.gov.au/parlInfo/search/display/display.w3p;query=Id%3A%22media%2Fpressrel%2F5375064%22).

Pearson, S., Honeywood, S., & O'Toole, M. (2005). Not yet learning for sustainability: The challenge of environmental education in a university. *International Research in Geographical & Environmental Education, 14*(3), 173–186.

Thomson, M. (2017). *The Cost of Defence*. ASPI Defence Budget Brief 2017–18. Canberra, Australian Strategic Policy Institute (ASPI).

White, H. (2012). *The China choice: Why America should share power*. Melbourne: Black Inc.

Chapter 2
Australian Contributions to the History of Military Geography

Stuart Pearson

1 Introduction

War is an intensely practical, human and physical activity affected by geography, history, weather and terrain (Smith 1990). The importance of geographic place, space, areas and concepts in military thought and practice makes an appreciation of the history of strategy and military geography fundamental. The evolution of military geography and its growth as a discipline is shaped by new values and technology. This chapter suggests that, so far, the themes, periods and individual cases in Australia's contributions to strategic and military geography have grown more like garlic than an onion. The cloves I consider in this metaphor formed around individuals and during periods of time in response to events and problems. The sub-discipline of military geography in Australia has a substantial tradition, a heritage of practitioners, scholars, and growing demand. It is surprising then that academic geography in Australia has contributed so little flavour to military geography. Internationally, military geography's epistemological and ontological questions are hotly contested; the sign of a healthy discipline.

Conventional military geography defined, 'as the application of geographical tools and techniques to the solution of military problems' (Galgano and Palka 2012) has a very long history in the discipline of geography. Maps are critical to the concepts and work of the military. Other spatial technology continues as a revolutionary driver in the military domain. Woodward (2016), in response to human geography's apparent disinterest in military issues in the post-war period, is leading a critical turn to define military geography narrowly as, "the study of the ways in which militarism and military activities are geographically constituted and expressed", but a

S. Pearson (✉)
School of Physical and Environmental Science, University of New South Wales,
Australian Defence Force Academy, Canberra 2610, ACT, Australia
e-mail: s.pearson@unsw.edu.au

S. Pearson et al. (eds.), *Australian Contributions to Strategic and Military Geography*, Advances in Military Geosciences,
https://doi.org/10.1007/978-3-319-73408-8_2

broader definition is appropriate and helpful – as illustrated in the chapters of this book.

In Australian geography the sub-discipline of military geography is very inconspicuous compared to the United States and British traditions. Australian universities' geography is driven by various factors (Holmes 2016) and a decline in integrated (human and biophysical) approaches such as regional geography, natural hazards or relevant methodological approaches (such as geographical information science) invites further concern about the contribution geography makes to strategic and military geography.

In 2015, when Geographers from universities, geo-intelligence company representatives, former and serving military officers and others gathered in a military geography session of the Institution of Australian Geographers conference it was a small group. They wanted to rediscover the history of Australian military and strategic geography and ensure they built on previous thinking. They wanted to avoid old mistakes and know the opportunities and threats of a new sub-discipline. Secondly, the group wanted to consider if strategic and military geography's past in Australia was an important resource for its possible futures. Also, during discussions with the American Association of Geographers military geography study group, it became clear there were some uniquely Australian contributions to strategic and military geography.

This is an initial scoping of Australian contributions to military geography and is not comprehensive or definitive; it is based on a review of the literature and from experience in military, educational and research efforts. It omits a discussion of school and university geography content (Hay 2012) and Australian School Cadets (but see Stockings 2007) as contributors to military geography. It is inspired by John Holmes's great insights into flood and ebbs in Australian geography and its sub-disciplines (Holmes 2002; Holmes 2014; Holmes 2016).

I asked venerable Professors of Geography for their considered opinions about the history of military geography and what they shared shows the subtle nature of an Australian contribution to strategic and military Geography. One said, "I think in the past Australian geography has been more about peace than war. That is exemplified early on by Griffith Taylor." Another had the opinion that Griffith Taylor had done more harm than good. With their suggestions in mind I dived into the literature. Subsequently, colleagues have patiently discussed whether individuals, themes and periods were *in* or *out* of the military geography domain and then whether these were *Australian* contributions or *strategic* in nature. I reiterate that this is an initial exploration and a provocation.

2 Synergies with Australian Military History Contributions

Australia's military history is well recorded and pervades Australian culture. History strongly contributes to strategic and geographic insights. The extraordinary Charles Bean (Bean 1983; Stanley 2017), the work of the Australian War Memorial and the support of an active cadre of quality historians (Grey 1999) results in Australia having a mature historiography (Holbrook 2014; Kent 1985) that has described the military experience and Wars, critiqued the silences of Frontier Wars (Reynolds 2006), the mythologising and commercialisation of wars, and commemoration (Stockings 2012). The discussions of the *landscapes of legends* (Stockings 2012) and what this means to the present and future (Brown 2014) show the mature characteristics of Australian military history with its strong geographic connections.

Monuments in Australian landscapes to war (Bulbeck 1991; Inglis and Brazier 2008) are imbued with strategic and geographical significance and these strongly contribute to the way contemporary military and political issues are framed (Ubayasiri 2015). This is evident at many scales. The military unit histories are woven with geographical insights; for example the Australian military surveyors (Coulthard-Clark 2000) and military lawyers (Oswald and Waddell 2014) commissioned histories are richly regional narratives. Even the Defence industry (Coulthard-Clark 2003) have commissioned their own histories that provide (sometimes selective) portals into the values, knowledge and the strategic contributions of these military groups. Military geography discovers its values in its history.

3 Military Geography in Early and Colonial Australia

People arriving in Australian waters more than 60,000 years ago were geographers by nature and their travel was strategically intentioned. These earliest Australians arrived at times and using technology that is the subject of ongoing research (Hiscock et al. 2016). More recently from 1642, Dutchman Abel Tasman was one of the first Europeans to contribute Australia's position: he mapped substantial portions of Australia, New Zealand and Pacific Islands into Europe's strategic and geographic knowledge. The exceptional chart and journal work of the British naval officer James Cook shows he was one of greatest-ever maritime explorers in the region (Beaglehole 1992). Later the phenomenal Matthew Flinders circumnavigated Australia (Flinders and Flannery 2001; Morgan 2016) and produced maps of enduring civil and military value (Kenshole 2017 this volume). Flinders contributed to making the title 'Australia' stick to the continent. A separate chapter in this volume plumbs the maritime geography further (Kenshole 2017).

On land many military-trained cartographers, most notably Thomas Mitchell (Baker 1997), reported geographic knowledge of Australia's land and people. The importance of these early maps to knowledge and visions of Empire in this period of colonialism can scarcely be exaggerated. Their records and geographical contribution was followed by settlement—settlers, soldiers, convicts and explorers. Soon the fears of colonialists triggered construction of forts (Burke 2008), indigenous Aborigines were dispossessed (Reynolds 2017 this volume), new tenure arrangements were made and the geospatial aspects of civilian and military geography spread. Throughout this process the mapping of changing interests continued. Australia's military geography still responds to changing domestic interests, values

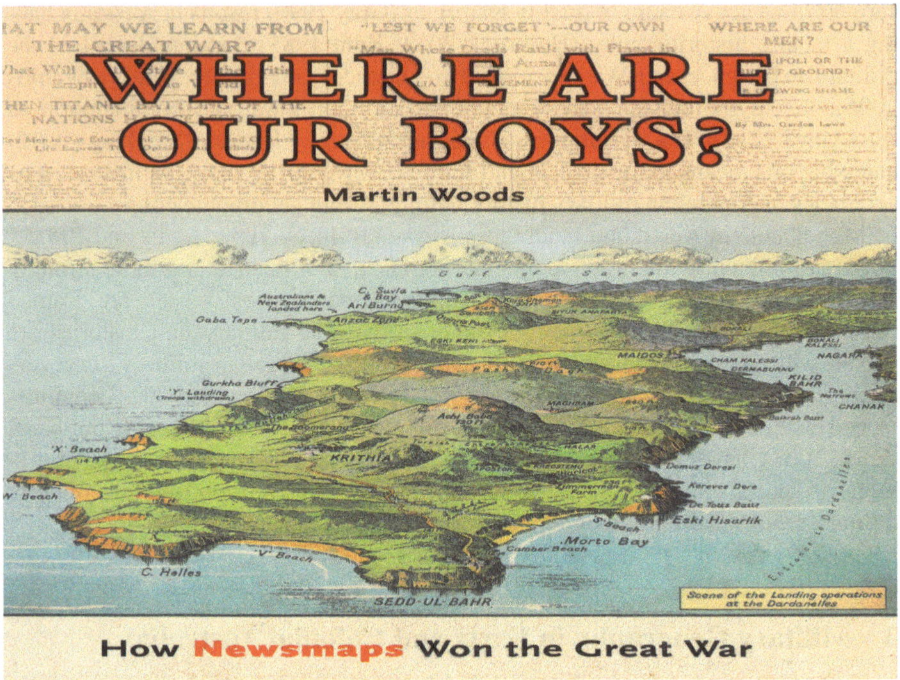

Fig. 2.1 The insatiable demand for newspaper maps of military topics during WW 1 is shown in the beautifully illustrated book by Martin Woods (2016). The cover image (a block diagram in the style made popular by Griffith Taylor) shows the Dardanelles—a landscape now imbued with so many meanings through military experience

and purposes. Maps are key artefacts for military geography. This strong historical tradition of mapping continues, as described in maritime maps (Kenshole 2017 this volume), Defence Estate management (Thackway and Ford 2017 this volume) and geospatial intelligence (Coorey 2017 this volume) and in the maps Australian's use to describe their strategic interests (Austin 2017; Kilcullen 2017; Press 2017- all in this volume).

The insatiable demand for newspaper maps of military topics during World War 1 (WW1) is illustrated in an excellent book by Martin Woods (2016) and its cover (Fig. 2.1) This remarkable collection demonstrates the demand and value of virtual military geography, communicated by map and media, to the people 'back home'. Australian's appetite for 'newsmaps' is a continuation of the use of maps as ways of sharing information, as icons and as tools of war.

The factors that stunted military geography in Canada, as explained by Jean Martin (2004), probably also apply in Australia: Canada was dominated by Imperial needs until the 1930s and therefore was not involved in world strategy or territorial ambition, and was isolated from threats of invasion. Under these circumstances Imperial Military Geography focused national military geographers on mapping and surveying that was largely done by engineers and delivering a narrow set of geographical insights. Jean found it difficult, despite the demand, to find an academically trained geographer or one in academia interested in military applications in Canada. She proposed a revival was necessary. The South African military geography experience is similar and they are also suggesting a revival is underway (Jacobs et al. 2002; Smit et al. 2016).

4 Australia's Endogenous Contribution

The rise of Australian strategic and military thinking that could be described as independent of the British Empire thinking grew after Federation in 1901 and the Defence Act 1903. Horner (1990) describes how two strategic views (the *Imperialists* and the *Australianists*) formed during this period and continued to generate what became known as Australia's strategic military gap that he observes today. Through overseas deployments alongside powerful allies the Australian Defence Force has developed a reputation for tactical excellence, however there have been few theatre-scale and political-scale opportunities for leadership and limited independent operational experience (Horner 1990). That has implications for the components and scale of military geography because the focus on tactics and operations defined the necessary set of geography skills. Strategic training has instead been managed by studies in the disciplines of history, politics and international relations.

In addition to the shortage of experienced strategic military leaders, Australia has a disdain for tall poppies in popular culture (Feather 1994) that has shaped the development of military geography. Australian identity is focused on the experience of Diggers - not the officers (Inglis and Brazier 2008). This was perhaps reinforced by Australia's suffering the grief of WW1 soon after Federation and Charles Bean's (Inglis 1970) deliberate war history approach that focused on the common soldier. This disdain carries into the focus of public monuments and sentiments to the Diggers. There are few officers with hero status: people like Sir John Monash (WW1 General) or "Weary" Dunlop (WW2 Medical Colonel) are rare in Australian popular culture (Dean 2010), and recognised for operational and humanitarian distinction rather than strategic achievement. Private John Simpson and his donkey, gathering wounded at Anzac Cove, has become an icon and Peter Cochrane's (2013) book explores the complexities of that icon's contribution to history, geography, politics and culture.

Most of the geographers I will mention here are much less well known than those leaders or historians. The past is being invigorated and reinterpreted through a growing remembrance industry that includes battlefield pilgrimages for youth, support for military service and overt nationalism (Brown 2014; Jennings et al. 2015) and it is unclear how this relates to the military geography needed for emerging Defence's roles in conflict prevention, change capacity development and reconstruction missions, social leadership and climate change (Boulton 2017, Thomas 2017).

Fig. 2.2 Griffith Taylor during the survey of the national capital on a horse, Canberra, 1913. (Source Professor Griffith Taylor collection, National Library of Australia)

It is useful to use some examples to explore the nature of strategic and military geography in Australia. I have chosen Griffith Taylor, Donald Thomson and the group of war-time Geographers to scope what an Australian contribution to military geography may involve.

5 Griffith Taylor

Thomas Griffith Taylor deserves a special place in any history of Australian Geography as founding Associate Professor of Geography at Sydney University Australia. His contribution is already well and graphically described by Carolyn Strange, if a little irreverently (Strange 2010; Strange and Bashford 2008). Griffith Taylor's career included frontier exploration of Antarctica, surveying Canberra (as it became the national capital, Fig. 2.2), two World Wars and their aftermath. Professor Tim Flannery, describes Griffith Taylor as "one of the greatest and most courageous scientists that Australia ever produced" (Strange and Bashford 2008).

Griffith Taylor's ambition, multi-disciplinary engagement (nurtured by Professor Edgeworth David, the Professor of Geology and Physical Geography, 1893–1924), pugnacious behaviour and sometimes overly-wide scope in life and work could personify Australia's strategic contributions to geography (Strange and Bashford 2008). However some things would be missing—things Griffith had but are currently missing in contemporary Australian Geography—strong regional insights, political and social engagement and grounded approaches (Carruthers 2009). Griffith Taylor's pyrolytic contributions to Australian population and sustainable agriculture debates, his demolition of racial pseudo-science, his position on environmental determinism and scientific national planning saw him lampooned by detractors.

Taylor's work in northern Australian for the Bureau of Meteorology, on request from Chief of General Staff Major General James Legge, triggered his interest in northern and central Australia and generated the research evidence he used against boosters of development potential in those areas. His participation in the debates of the period was sensational. His opposition to the White Australia policy, based on anthropo-geographical evidence and his rejection of racial differences through empirical analysis (*Environment and Race* 1927), are unequalled in Australia. Using his press-ready personality and popular promotion of Antarctic exploration, he delivered superior evidence-based arguments against predictions of 'Australia Unlimited' whilst inspiring public interest in scientific endeavor. He attacked the unrealistic population predictions of the boosters and cemented his contribution in Australian geography. However Taylor made enemies and his environmental possibilism was too close to environmental determinism for the social geographers of the post-war period. He, and geography generally, was tarnished by association with Karl Haushofer, the geographer who informed Nazi policy-makers (Strange and Bashford 2008).

Griffith Taylor was not a military man but he was supportive during WW2 and active in planning the peace (Strange and Bashford 2008). After the war, Griffith Taylor espoused Geopacifics to his profession in what he described as, "an attempt to base the teachings of freedom and humanity upon real geographical deductions; it is humanized Geopolitics" (Taylor 1946). He proposed a rational set of crop and power-related regions to prevent conflict in Europe by ensuring that neither race nor nations would cause conflicts and proposed that "we should study environmental control so as to advance in harmony with our environment" (Taylor 1946).

In his presidential speech to the Institute of Australian Geographers titled 'Geographers and World Peace: A Plea for Geopacifics' (Taylor 1963) he said,

> …so few geographers are prepared to explore public problems which touch on geography as much as on most other disciplines. This is certainly the case with the problem of world peace."

And he went on to recommend,

> "that all university students are exposed to a course of lectures on civilization, preferably of an objective graphic type such as I have tried to offer in my studies of geopacifics. If this type of scientifically-derived data is grafted on to the age-long teachings of religion, we shall, I hope make real progress towards a world at peace." (Taylor 1963)

Griffith Taylor edited the magisterial book, *Geography in the Twentieth Century* (1951), that included a dendrogram showing the many branches of Geography (on which there was no 'military geography'). He had predicted the future settlement pattern of Australia in 1919 and 30 years later could marvel at his prescience (Taylor 1951). His emphasis on futures linked to his scientific determinism (Taylor 1951:12), and his work in Antarctica and deserts was repeatedly used to show the importance of environmental considerations in considering the management of marginal lands.

Taylor (1951:18) described Geography as a 'liaison subject'; with geographers working as "buccaneers or pirates with other sciences and philosophy". He promoted himself, geography and science relentlessly for the betterment of civilisation, nations and citizens (Spate 1978; Strange 2010; Strange 2012; Strange and Bashford 2008) and is an example of a practitioner of strategic and military geography thinking.

6 Donald Thomson

In contrast to Griffith Taylor's high profile, Donald Thomson is best known for his peace-making mission to Caledon Bay 1932–3 and organisation of Yolngu Aborigines to act as defenders during WW2 (Rigsby and Peterson 2005; Thomson and Peterson 1983). The Australian National Research Council funded his (1928) anthropological and zoological work on Cape York Peninsula, Queensland, for which he was awarded the patron's medal of the Royal Geographical Society (London), it had also prepared him for what his academic colleagues described as his role as 'Australia's Lawrence of Arabia' (Morphy 2002). He worked with local warriors in WW2, later advocating Aboriginal recognition and opposing forced relocations of desert Aboriginal people for the Woomera Test Range in the Cold War (Morphy 2002; Morton 1989). According to Museums Victoria, his UNESCO listed collection is one of the most comprehensive and significant collections of Aboriginal cultural heritage material in the world. Donald Thompson and the warriors have a complex legacy in the Australian Defence Force and there is a rich Australian Aboriginal and Torres Strait Island service history (Riseman 2013) worthy of further research and understanding (Reynolds 2017).

7 Wartime Military Geographers

The Allied Geographical Section in the Southwest Pacific Area generated a massive amount of photography, topographic mapping and a library of restricted evidence for military operations in the Pacific. Suitably labelled to acknowledge their potential usefulness to the enemy if captured, these included a military purposed regional geography and detailed aerial photograph interpretations to inform operational decisions. The knowledge, skills and insights developed during this period provided a valuable source for future research (Bruce Thom, pers. comm 2017), a broad scale application of remote sensing and image technology and insights into the importance of geographic information (Coorey 2017 this volume). The war-time geographical work is a resource for further research and operational information—for example, the contributions of Rhodes Fairbridge and Trevor Langford-Smith in mapping the Pacific Islands remains an unexplored source for geographers. The collection of maps, photos and pamphlets at the Australian War Memorial is an opportunity for historical geography to be applied to the products and the people involved in this massive war-time effort. It would also provide materials for comparative and trend analysis in geography.

8 Post-WW2 Contributions

The demobilisation and return of military service personnel to civilian life in Australia had massive social and environmental consequences. Space precludes discussion of service-related settlement schemes in Australia (Prescott 1969). These legal controls on property and land use continue to shape domestic and international strategic possibilities (Powell 1988) and further research on this is needed. Those Australians returning from service abroad were changed forever; they had participated in a multinational mobilisation of society, industry and science to a single task. For many people this included learning new geographical skills, awareness and knowledge. The atomic bomb, and the successful strategic application of the nation and scientific methods to war, provided strong domestic support for peace, science and planning. Resources were directed into systematic surveys of the land resources and preparing for the future development of Northern Australian through integrated assessments of existing uses, topography, vegetation and soils (Fig. 2.3). The underlying factors, such as climate, geology and geomorphology, were mapped at a scale appropriate for policy-makers. These national survey groups, informed by air photography and coordinated by the government's central science organisation between 1946 and 1977, equipped with demobilised resources (Fig. 2.4). used the geographical methods, and even the block diagrams of Griffith Taylor (Fig. 2.5), to integrate information into a visual representation of regional geography. This employed people who went on to become the great names in Australian geography such as CR Twidale, Rob Galloway, Henry Nix, MAJ Williams and Jack Mabbut. According to Bruce Thom, this was a rare example of close cooperation between the academic researchers and national development. This work was later useful for planning biosecurity measures associated with bovine tuberculosis (Laut et al. 1985).

Taylor also stands as Australia's most prolific graphical geographer—applying his sketching skill to help visualise a full range of geographic phenomena including block diagrams (Fig. 2.5), that have impacts beyond the discipline.

The post-war period marked a massive Commonwealth Scientific and Industrial Research (CSIR) and university investment in science with rising student enrolment and fee income (Forsyth 2017) generated a well-educated and technically advanced society influenced by Taylor's ideas. For scientists (and geographers) WW2 had shown the power of science purposefully applied and the dangers of war and industrialism. Sir David Rivett (quoted by Deery and Clohesy 2013), chief executive officer of CSIR warned, even before the end of WW2, in a keynote address to scientists:

> "If I am not mistaken, there is a fight ahead in Australia (and not only in Australia) to keep Science as the pioneer, the discoverer, the leader, and not just the patronized servant called in at times of distress and low dividends to act as the clever artisan to keep the ship of Industry in motion (or to keep the dogs of war supplied)."

This was a heady time for Australians; there was a mood for global change and internationalism. In addition to Griffith Taylor's Geopacifics and other internation-

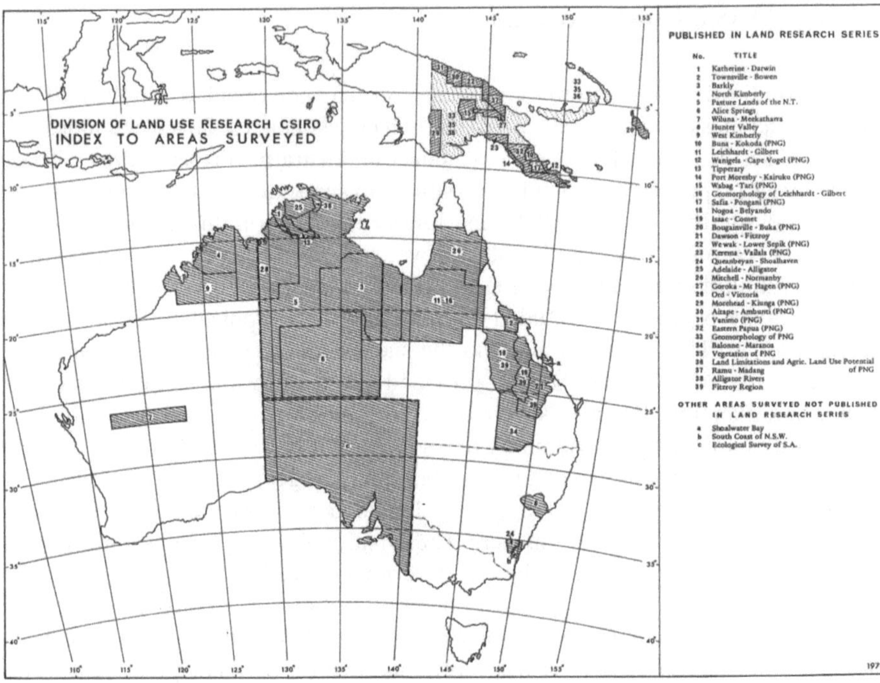

Fig. 2.3 The CSIRO's Land Research Series delivered 39 reports on parts of Australia and Papua New Guinea by teams of researchers applying geographical methods. (Source CSIRO 1946–1977)

alist suggestions previously described, "Doc Evatt", the Australian Attorney-General and Minister for External Affairs, became the President of the United Nations General Assembly from 1948 to 1949 with a strong global humanitarian agenda. He helped draft the Universal Declaration of Human Rights, negotiated the creation of Israel and supported the League of Nations—the rules-based global order that Australia espouses (DoD 2016b) was set-up during this time and strengthened over the next seventy years.

9 Cold War Australian Contributions

Yet by 1948, the Cold War, atomic and weapons secrecy, and espionage arrests, put fears of internationalists or communist scientists into the minds of political and security services. The science community, "deserted the firing line for the safer territory of political neutrality" (Deery and Clohesy 2013). Academics, publications and their work, like Griffith Taylor (1951) separated themselves from national security and directed the geographical discipline away from military geography. Even though Australia's experience of this Cold War and anti-communist split between Universities and the state's security and military (Deery and Clohesy 2013) was

Fig. 2.4 Demobilised World War resources applied to national survey science. Survey vehicles near Anthony's lagoon, Barkly Region Northern Territory–Queensland 1948; the left-most one is a "Blitz Truck" and is still seen in Australian paddocks and sheds. Known commonly as the "Blitz" it was officially the Canadian Military Pattern (vehicle). At the other end is the US-style jeep and trailer. A Chevrolet 12 Cwt General Service Utility and Dodge WC51 sit between them. The expeditioners wear ex-military garb. (Source CSIRO 1946–1977)

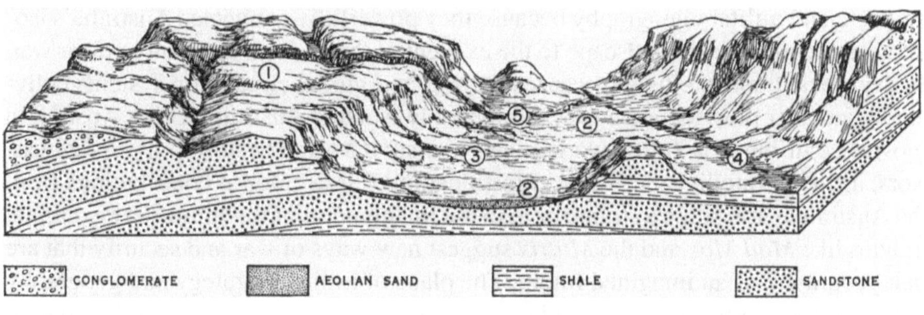

Fig. 2.5 The block diagrams used in CSIRO's post-war reports were a continuation of Griffith Taylor's tradition in graphically illustrating integrated landscape approaches. For an explanation of the labels see the original publication. (Source CSIRO 1946–1977)

weaker than in the United States, it cast a long shadow (Coulthard-Clark 2003). In a way, when the Australian Prime Minister in his 2016 Defence White Paper increased funding of Defence-Industry-Research collaborations, it was continuing efforts to repair a split between military and civil research that is decades old.

Australia's shifting strategic alliance to the United States (from the United Kingdom) after WW2 is worthy of further research from the perspective of military geography—Australian military doctrine, equipment and deployments are now closely aligned with the United States. The realignment of geography to the United States traditions was also marked during this period. Bruce Thom, in the Faculty of Military Studies (UNSW) Duntroon in the 70s, contrasted US Office of Naval Research (ONR) interest in research, including expeditions to the Australian coasts and the Ord, with the lack of interest of Australian Defence (Thom 2017). The US Geography Branch was interested in developing a broad understanding of coastal environments, processes and understanding what had not worked in coastal operations during WW2. Military support for US professors and graduate students in contrast to Australian academics experience with the Australian military made clear the problem Australians had linking military people to research. In addition, nuclear missiles and other technology challenged geography's privileged position in military doctrine (like it did to history (Evans 1990)) because military thinking was focused on global and instantaneous war. This fostered a new discipline of strategic studies with its grand challenges of escalation scenarios played at global scale. The Cold War gave way to managerial approaches in the Australian Defence Force in the 1960s and the counter-reaction in the 1980s as regional, expeditionary and persistent wars as ally to the United States continued into the twenty-first century.

Perhaps tangentially, but illustrative of the importance of reframing strategic and military geography on cultural scales, the interaction of global and individuals during the Cold War shows how useful cultural, social and agent-based approaches could be to explaining developments in military geography. The Cold War and fear of nuclear war was a paradigm shift in culture and science with important implications. The science fiction book *On the Beach* (Shute 1957) and movies (1959 and 2000), in ways too interesting to summarise briefly, were Australian contributions to strategic and military geography because they powerfully connected Australia's isolated and iconic beach landscape to the existential threat from a distant nuclear war. In the movie poor intelligence causes the war and global circulation systems delivers the nuclear oblivion even to geographically isolated Australia. Other Australian movies contributed to future world views that are dark and deadly; full of road warriors, artificial intelligence (AI) and fortification. These cultural contributions from the Australian arts helps to frame or map the imagination of the future. For example, movies like *Mad Max* and the *Matrix* suggest new ways of war and security that are in a geography of an imagined future. The place of military strategy and geography is haunted not only by history and experience but by our selected imaginations of things worth protecting or fearing in the future (Boulton 2017 this volume).

In fact, it was *On the Beach* that motivated Australian pediatrician Helen Caldicott—the anti-nuclear warrior named by The Smithsonian Institution as one of the most influential women of the twentieth century—in her work against radioactive, nuclear and space weapons. She was a catalyst for decades of anti-nuclear actions that stopped French atomic tests in the Pacific and held one of Australia's key alliance (ANZUS) at risk (Carpenter 1986; Pugh 1989) and thus shaped policies that, until recently, stopped United States warship visits to New Zealand. Her sustained criticism of United States military exceptionalism and assumptions of limited nuclear war has contributed to Australia's position on nuclear weapons. The role of issues raised in the media resonating with the public and then influencing strategic and military geography is worthy of further analysis.

In retrospect the end of the Cold War and the associated treaties (Comprehensive [nuclear] Test Ban Treaty, Chemical Weapons Treaty, Missile Technology Control Regime (MTCR)) were a brief lull. The current continuous war narrative—regional wars, non-state wars, star wars, wars on terror, drone wars and cyber wars have eroded the faith people placed in the United Nations and in the prospect of permanent peace. The geography of the contemporary period is complex: the rise of regional sub-national conflicts; the widening obligations to protect and intervene in civil disputes; the conflicts of the Middle East; third party conflict spaces; redefining the new world order; the decline and rise of Russian-US tensions; sustained China-US tensions; periods of regional neighbour instability; home-grown terrorism; natural hazards and disasters; Timor-Leste's post-independence challenges; and a weakening PNG all lend themselves to a stronger geographical approach. There has not been a more appropriate time for the spatial, temporal and scalar skills, the geographical approaches to human and biophysical knowledge and attitudes to integrated problem solving to reframe the problems and solutions.

10 The Contemporary Era

The national security and military assessments of the contemporary era suggests the need for resilience, allies and good intelligence (DoD 2016b). Australia contributes to a network of places for multinational security information exchange (Pine Gap, Nurrungar, North West Cape, Darwin) that, with the alliances of ANZUS (Australia, New Zealand and the United States) and Five Eyes (adding Canada and the United Kingdom), make it possible to gather high quality information at an acceptable cost. This service makes the United States, according to Dibb's remarkable access and insights, an irreplaceable ally (Ayson 2016) with strategic and military consequences. However, Helen Caldicott, and others, are adamant that the military-industrial complex (first used as a cautionary term by Eisenhower in 1961), strengthened by policy and investment (DWP 2016), is to be feared and resisted. Most Australians see the alliance with the United States in a positive light. Australia can as an ally, according to the now United States Defense Secretary James Mattis, contribute advice and be a valuable moderating influence.

Nearly a generation after WW2, Australian barrister Geoffrey Robertson's work on war crimes and human rights (2006) contributed to changing thinking about military humanitarianism—and the complexities of tasking militaries to deliver humanitarian goals and prevent genocide (Chandler 2001). Much more controversially, the Australian cyber-activist Julian Assange contributed to rethinking ethics and the conduct of security and military options at the global scale through the Wikileaks release of Iraq and Afghanistan Wars and other documents (Assange 2015; Fowler 2012).

Australia has deliberately supported the policy development and then implementation of the United Nations Security Council's Resolution 1325 on Women, Peace and Security. Australia has been a strong and consistent advocate of this through its foreign policy, aid, police and military programs (Australian Government 2012). The Australian Defence Force continues to deal with the mistreatment of women within the military, cultural change initiatives (this is ongoing and most clearly enunciated by then Chief of Army Lieutenant General David Morrison in 2013) and the removal of gender restrictions from ADF combat roles in 2011 and again in 2016. The depth, sophistication and potential benefits of gender mainstreaming in the Australian Army is remarkable (Boulton 2017). There is a thread joining these; Australians in the contemporary era are making important contributions to geography in strategic and military terms.

10.1 Geography's Contribution to Educating Military Officers

Geography discipline leaders (including Associate Professor Eric Woolmington, then Professors Bruce Thom, Roger Mclean and Brian Lees) with specific and direct responsibilities for educating Australian military officers as Geographers at Royal Military College (Duntroon prior to 1985) and then at the Australian Defence Force Academy, all pursued a liberal educational philosophy rather than a military-aligned one. This was done with the support of the military leadership of the Academy to honour the agreement between the University and Defence.

As Horner (1990) highlights, and Kilcullen (2010) experienced, an Australian officer today needs to be better educated, more strategically aware and understand geography better than WW1 officers who may have had one thousand times more people at their command. Like others (O'Malley 2010), the Australian Defence Force is preparing to be resilient in a complex and uncertain world. Geography as a discipline contributes to that preparation. Australians have a sustained claim on the concept of resilience in science (Walker and Salt 2012) and a culture of military resilience (Gilmore 2016; Kilcullen 2007) defined by historical evidence. Defence understands that improving resilience and innovation will be of ongoing importance under climate change and other uncertain future conditions (DoD 2016b).

10.2 Contributions to Counter Insurgency

Some of Australia's officers have been feted as innovators. For example, tunnellers trained as geologists by Sydney University's Professor Edgeworth David are lauded for their innovations in trench-warfare, and Gen Sir John Monash demonstrated mastery of integrated planning, executing complex logistics and innovative rolling artillery barrages in WW1 (Serle 1982). Others, such as the much less well known, Brigadier Ted Serong, redeveloped Australian jungle training and wrote widely on counter-insurgency, filling advisory roles in South Vietnam and later consulting to the United States where Australian military contributions to operations and tactics continue to resonate (Kilcullen 2017).

The international deployment of military force to counter insurgents and failed-state interventions has grown in frequency and scale since the Cold War. These conflicts are in full and fast view of the global media and, coupled with an acute ethical awareness, means that every action may have geopolitical consequences. David Kilcullen has contributed to the innovative and holistic thinking needed for successful counter-insurgency. His remarkable contribution includes his tactical experience—such as a Timor-Leste border misunderstanding that involved shots being fired before quick thinking, good language skills and shared complaints about poor quality maps restored order (Kilcullen 2010). He acted as a catalyst within Defence and subsequently with media and NGOs to achieve change in the ways Middle East conflict and counter-insurgency are understood in doctrine. Whether working as an Australian advisor to General Petraeus, as an author of counter-insurgency manuals or highly accessible books and newspapers, Kilcullen contributed strategically to changing the way the United States and Australian military fights. Geography academics from West Point agree his contributions have been significant—he is well known. The timing has been important because United States senior leaders like Mattis and Petraeus themselves have been described by Kilcullen as "an insurgency within the bureaucracy" that is driving military innovation (McLeod 2017) in the strategic and military interventions required for the new challenges. David Kilcullen's recent work observes littoral urbanisation as one of the most important mega-trends of the twenty-first century and then raises concerns and possible strategic countermeasures for the emergence of conflict where urban governance breaks-down (Kilcullen 2015).

10.3 Strategic Geography and Geopolitics

In the final section it is important to briefly review the discussion between military and geopolitical policy makers. When Paul Dibb (2006) asked "Is strategic geography relevant to Australia's current defence policy?" he asked it from the pages of the *Australian Journal of International Affairs* and it was subsequently ignored by geography journals. He had previously posed this view in his advice to the Minister in 1986. The nexus, he argued, between strategic geography and force structure had been broken as Australia had developed an expeditionary force to act in subordinate roles alongside the US and allies in distant theatres of operations. He argued that Australia should apply strategic geography as an 'iron discipline' (Dibb 2006) and continued to use maps to show the geographical nature of Australia's interests—including a map from Griffith Taylor to show the durability of this approach (Fig. 2.6).

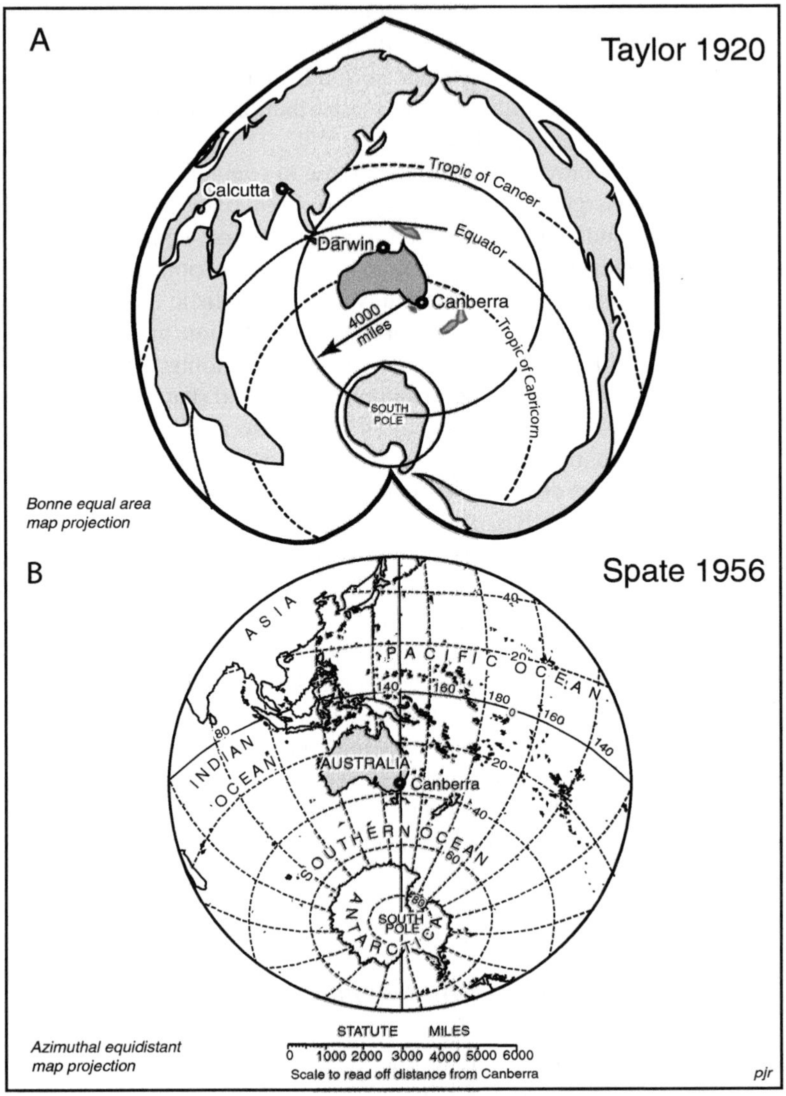

Fig. 2.6 This map epitomises the geographers' approach and shows continuity of focus on spatial issues between Griffith Taylor's 1920 and Spate's 1956 maps. Map A shows the great ocean's isolation of Australia, with a radius of 4000 miles from Canberra shown to pinpoint stepping stones through the Dutch East Indies for flights to England via Calcutta. Map B shows the hemisphere around Australia centred on Canberra—all points on the map are at a proportionately correct distance from the centre point. (Source Rimmer and Ward 2016)

Fig. 2.7 Australia Defence Force Operations during 2015–2016 according to the Defence Annual Report. (Source DoD 2016a)

Dibb's substantial contribution to strategic and military geography has been expertly described in the multi-authored book and it surfaces in his criticism of Alan Dupont's narrative about the global reach of Australian interests. Commitments to expeditionary forces in subordinate roles to allies is the character of Australia's strategic and military deployments—as shown on the ADF map of operations during 2015–16 (Fig. 2.7).

Dibb, as a geographical possibilist, puts geography explicitly into the analysis whilst also observing that the pattern of past and current deployments of the ADF (Fig. 2.7) does not replace the fundamental role of the ADF. There is an echo of Griffith Taylor's contrarian and evidence-based style when Dibb presents his view that the muddled strategies of policy-makers relate to changing United States coalition requirements on which Australian strategic interests and capabilities have been locked. The 'geographical force fields' of Dibb and others (Ayson 2007) involve knowing the strategic importance of the immediate neighbourhood, interests in the broader Southeast Asia region, then in Northeast Asia, and finally global strategic interests. Dibb argues that geography and resources should focus preparedness for wars of necessity (territorial threats and in the immediate region) rather than wars of discretion (peace keeping and coalition operations). He is skeptical that the expeditionary force in being is truly designed for 'Australian interests', in that it includes the recent acquisition of two of the largest amphibious ships in the region and a highly technical and United States dependent force.

Mark O'Niell, Alan Dupont and others (Kilcullen 2017 this volume) criticised Dibb for clutching to the 'Defence of Australia' and strategic geography case because it ignores the actual ADF tasks (Ayson 2016) and it overstates the local and regional interests of Australia. Experience certainly suggests failed states, disasters, United Nations interventions, working for stability in our region or attempting to destroy a support base for Islamic terrorism are the new challenges for the ADF. However, the future looks quite different and the discussions are an opportunity for geographers. There are many challenges that strategic and military geography could inform. New knowledge of the issues of megacities, the rise of China, India and Indonesia as trading partners or threats and the growing pressures of regional superpowers is needed. As circumstances change the discussions of Australia's strategic interests also adjust to the rise of China (White 2012), India (Hall 2016) and Indonesia that are described individually and within geographical concepts such as regions and maps (even when neither are provided!) (Medcalf 2014). Geographers, and specifically those with strategic and military interests, have been in very short supply in this dynamic environment.

11 The Current Turn to Strategic and Military Geography in Defence

A small group of strategic and military geographers are very active in developing systems and holistic approaches to the research and development needs of Defence. Working within the Defence Science and Technology Group (DST) and in the Department of Defence, people provided the capability within the ADF to organise analysis and awareness raising activities of the likely impacts of global changes in strategic military geography (Holloway 2017 this volume). This effort tapped into global academic work on the Anthropocene and complex, interacting changes in biophysical, human and cyber geographies. That innovative work was sponsored by the Vice Chief of the Defence Force Group and used strategic military geography to articulate plausible threats Defence could be called on to meet, and the operating contexts in which these might need to be addressed. Using energy, climate change, space weather and chemical loadings and thresholds in the Earth's system as foci and testing them through a series of stakeholder workshops, it was climate change that was short-listed as having a major effect in the future. People involved in ADF Preparedness demanded more information on how, where, and how often Defence could be tasked to operations such as peacekeeping; humanitarian assistance and disaster relief; as well as domestic support and border protection operations. These discussions draw on military, scientific and civil society, and provide potential for a wider and more diverse interaction. There is now more demand than Prof Bruce Thom found in the 1970s for actionable knowledge.

The international context for Australian military geography has changed; people are surprised at the advanced considerations of the other militaries, the new expected tasks of the military, and the growing gap between the challenges and the responses. The depth and capabilities of the national security agencies, including Defence, to consider the longer-term, the uncertainties and the risks has developed remarkably since the 1970s. They share with geographers a commitment to integrated and systemic approaches to exploring uncertainty and mapping complexity to achieve more desirable futures. Between 2015 and 2017 the sub-discipline of military geography presented at domestic, international and interdisciplinary conferences and there are strong opportunities to revitalise military geography as a concept, and as a sub-discipline and to carry ideas into decision-making.

12 Conclusion

This is, necessarily, a qualified introduction to the Australian contributions of strategic and military geography. It seeks to build and broaden interest, to facilitate discovery, and provoke new engagement. With advice from the few experts, and a plunge into the literature and archives, this chapter has identified important links to military history, opportunities concerning boundaries, legal, gender and regional geography that are likely to be fruitful domains for those interested in military geography. There are more opportunities and threats than there are people to think about them. Some, such as strategic futures and planetary boundaries that "must not be transgressed" (Rockström et al. 2009), need our urgent attention and that existential threat cannot be left to the military alone. Purposeful geography, that integrates human and physical geography to generate new skills, knowledge and attitudes, represents a continuation of the history of Australia's contribution to military geography that has existed in the past and is needed for the future.

References

Assange, J. (2015). *Introduction: WikiLeaks and empire The WikiLeaks files: The world according to US empire* (pp. 1–19). London: Verso Books.

Australian Government. (2012). *Australian National Action Plan on Women, Peace and Security 2012–2018*. Canberra: Department of Families Housing, Community Services and Indigenous Affairs.

Ayson, R. (2007). The 'arc of instability'and Australia's strategic policy. *Australian Journal of International Affairs, 61*, 215–231.

Ayson, R. (2016). The importance of geography. In D. Ball & S. Lee (Eds.), *Geography, power, strategy and defence policy: Essays in honour of Paul Dibb.* (pp. 71–83). Canberra: Australian National University.

Baker, D. W. A. (1997). *The civilised surveyor: Thomas Mitchell and Australian Aborigines*. Melbourne: Melbourne University.

Beaglehole, J. C. (1992). *The life of Captain James Cook*. London: Stanford University Press.

Bean, C. E. W. (1983). *The official history of Australia in the war of 1914–1918: The Australian Imperial Force during the Allied offensive, 1918* (Vol. 6). Brisbane: Univ of Queensland Pr.

Boulton, E. G. (2017). *Teaming: An introduction to gender studies, unshackling human talent and optimising military capability for the coming era of equality: 2020 to 2050*. Canberra: Australian Army.

Brown, J. (2014). *Anzac's long shadow: The cost of our national obsession*. Melbourne: Black.

Bulbeck, C. (1991). Aborigines, memorials and the history of the frontier. *Australian Historical Studies, 24*, 168–178.
Burke, A. (2008). *Fear of security: Australia's invasion anxiety*. Cambridge University Press.
Carpenter T. G. (1986). Pursuing a strategic divorce: The US and the ANZUS alliance. *Cato Institute Policy Analysis* (67).
Carruthers, J. (2009). Carolyn strange and Alison Bashford: GriffithTaylor:Visionary, environmentalist, explorer. *Historical Records of Australian Science, 20*, 135–137.
Chandler, D. G. (2001). The road to military humanitarianism: How the human rights NGOs shaped a new humanitarian agenda. *Human Rights Quarterly, 23*, 678–700.
Cochrane, P. (2013). *Simpson and the donkey: The making of a legend.* Carlton: Melbourne University Publishing.
Coulthard-Clark, C. D. (2000). *Australia's military map-makers: The Royal Australian Survey Corps 1915–1996*. Oxford: Oxford University Press.
Coulthard-Clark, C. (2003). *Breaking free: Transforming Australia's defence industry*. Melbourne: Australian Scholarly Publishing.
CSIRO (1946-1977) CSIRO land research surveys – on line. Commonwealth Scientific and Industrial Research Organization. http://www.publish.csiro.au/cr. Accessed 14 May 2017.
Dean, P. (2010). Commemoration, memory, and forgotten histories: The complexity and limitations of Australian Army biography. *War & Society, 29*, 118–136.
Deery, P., & Clohesy, L. (2013). 'Patronised servants': Australian scientists in the 1940s. *Journal of the Royal Australian Historical Society, 99*, 114–132.
Dibb, P. (2006). Is strategic geography relevant to Australia's current defence policy? *Australian Journal of International Affairs, 60*, 247–264.
DoD (2016a) Defence Annual Report 2015–16. Department of Defence, Canberra.
DoD (2016b) Defence White Paper. Department of Defence, Canberra.
Evans, M. (1990). Military history in the education of western army officers. In H. Smith (Ed.), *Preparing future leaders: Officer education and training for the twenty-first century* (pp. 121–143). Canberra: Australian Defence Studies Centre.
Feather, N. T. (1994). Attitudes toward high achievers and reactions to their fall: Theory and research concerning tall poppies. *Advances in Experimental Social Psychology, 26*, 1–73.
Flinders, M., & Flannery, T. F. (2001). *Terra Australis: Matthew Flinders' great adventures in the circumnavigation of Australia*. Melbourne: Text Publishing.
Forsyth, H. (2017). Post-war political economics and the growth of Australian university research, c. 1945–1965. *History of Education Review, 46*, 15–32.
Fowler, A. (2012). *The most dangerous man in the world: The inside story on Julian Assange and WikiLeaks*. Melbourne: Melbourne University Press.
Galgano, F., & Palka, E. J. (2012). *Modern military geography*. New York: Routledge.
Gilmore, P. W. G. (2016). *Leading a resilient force: Insights of an Australian General*, Canberra: Army Research Paper No 11.
Grey, J. (1999). *A military history of Australia.* Port Melbourne: Cambridge University Press.
Hall, I. (2016). India in Australia's 2016 Defence White Paper. *Security Challenges, 12*, 181–185.
Hay, I. (2012). Over the threshold—Setting minimum learning outcomes (benchmarks) for undergraduate geography majors in Australian universities. *Journal of Geography in Higher Education, 36*, 481–498.
Hiscock, P., O'Connor, S., Balme, J., & Maloney, T. (2016). World's earliest ground-edge axe production coincides with human colonisation of Australia. *Australian Archaeology, 82*, 2–11.
Holbrook, C. (2014). *Historiography 1918-Today (Australia).* Freie Universität Berlin. http://encyclopedia.1914-1918-online.net/pdf/1914-1918-Online-historiography_1918-today_australia-2014-12-16.pdf. Accessed 9 Mar 2017.
Holmes, J. (2002). Geography's emerging cross-disciplinary links: Process, causes, outcomes and challenges. *Geographical Research, 40*, 2–20.
Holmes, J. (2014). Explorations in Australian legal geography: The evolution of lease tenures as policy instruments. *Geographical Research, 52*, 411–429.

Holmes, J. (2016). Whither geography? A response to Finlayson's concerns. *Geographical Research, 54*, 103–106.
Horner, D. (1990). Australian Army leadership: The historical background. In H. Smith (Ed.), *Preparing future leaders: Officer education and training for the twenty-first century* (pp. 83–106). Canberra: Australian Defence Studies Centre.
Inglis, K. S. (1970). *CEW Bean, Australian historian*. Brisbane: University of Queensland.
Inglis, K. S., & Brazier, J. (2008). *Sacred places: War memorials in the Australian landscape*. Melbourne: The Miegunyah Press.
Jacobs, J. A., Van Rensburg, H. S. J., & Smit, H. A. P. (2002). Military geography in South Africa at the Dawn of the 21st century. *South African Geographical Journal, 84*, 195–198.
Jennings, P., Davies, A., Frühling, S., Goldrick, J., Kalms, M., & Medcalf, R. (2015). *Guarding against uncertainty: Australian attitudes to Defence 2015*. Canberra: Department of Defence.
Kent, D. A. (1985). The Anzac book and the Anzac legend: CEW Bean as editor and image-maker. *Australian Historical Studies, 21*, 376–390.
Kilcullen, D. (2007). Australian statecraft: The challenge of aligning policy with strategic culture. *Security Challenges, 3*, 45–65.
Kilcullen, D. (2010). *Counterinsurgency*. Oxford: Oxford University Press.
Kilcullen, D. (2015). *Out of the mountains: The coming age of the urban guerrilla*. Oxford: Oxford University Press.
Laut, P., Nanninga, P., M. & CSIRO Division of Water and Land Resources. (1985). *Landscape data for cattle disease eradication in northern Australia*. Canberra: Commonwealth Scientific and Industrial Research Organization.
Martin, J. (2004). Canadian military geography 1867–2002. In *Studies in military geography and geology* (pp. 53–64). Dordrecht: Springer.
McLeod, T. (2017). *New US secretary of Defence James Mattis and the lessons of history*. Melbourne: Centre for Policy Development. https://cpd.org.au/2017/01/travers-mcleod-on-us-secretary-of-defence-james-mattis/
Medcalf, R. (2014). In defence of the Indo-Pacific: Australia's new strategic map. *Australian Journal of International Affairs, 68*, 470–483.
Morgan, K. (2016). The making of a maritime explorer: The early career of Matthew Flinders. *Journal for Maritime Research, 18*, 1–16.
Morphy, H. (2002). *Thomson, Donald Finlay Fergusson (1901–1970)*. National Centre of Biography, Australian National University. http://adb.anu.edu.au/biography/thomson-donald-finlay-fergusson-11851/text21213. Accessed 5 May 2017.
Morton, P. (1989). *Fire across the desert: Woomera and the Anglo-Australian joint project 1946–1980*. Canberra: Australian Government Printing Service.
O'Malley, P. (2010). Resilient subjects: Uncertainty, warfare and liberalism. *Economy and Society, 39*, 488–509. https://doi.org/10.1080/03085147.2010.510681.
Oswald, B., & Waddell, J. (2014). *Justice in arms: Military lawyers in the Australian Army's first hundred years*. Sydney: Big Sky Publishing.
Powell, J. M. (1988). *An historical geography of modern Australia: The restive fringe, Cambridge Studies in Historical Geography* (Vol. 11). Cambridge: Cambridge University Press.
Prescott, J. R. V. (1969). *The geography of state policies*. London: Hutchinson.
Pugh, M. (1989). *The ANZUS crisis, nuclear visiting and deterrence, Cambridge Studies in Inernational Relations* (Vol. 4). Cambridge: Cambridge University Press.
Reynolds, H. (2006). *The other side of the frontier: Aboriginal resistance to the European invasion of Australia*. Sydney: UNSW Press.
Rigsby, B., & Peterson, N. (2005). *Donald Thomson: The man and scholar*. Canberra: Academy of the Social Sciences in Australia.
Rimmer, P. J., & Ward, R. G. (2016). The power of geography. In D. Ball & S. Lee (Eds.), *Geography, power, strategy and Defence policy: Essays in honour of Paul Dibb* (pp. 45–69). Canberra: ANU Press.

Riseman, N. (2013). Serving their country: A short history of Aboriginal and Torres Strait islander service in the Australian Army. *Australian Army Journal, 10*, 11–22.
Robertson, G. (2006). *Crimes against humanity: The struggle for global justice*. London: Penguin.
Rockström, J., et al. (2009). A safe operating space for humanity. *Nature, 461*, 472–475.
Serle, G. (1982). *John Monash: A biography*. Melbourne: University Press.
Shute, N. (1957). *On the Beach*. New York: William Morrow & Company, Inc.
Smit, H., Magagula, H., & Flügel, T. (2016). South African military geography: Advancing from the trenches. *South African Geographical Journal, 98*, 417–427.
Smith, H. (1990). The education of future military leaders. In H. Smith (Ed.), *Preparing future leaders: Officer education and training for the twenty-first century* (pp. 145–169). Canberra: Australian Defence Studies Centre.
Spate, O. (1978). Palaeoclimates of geographical thought. *The Australian Geographer, 14*, 1–7.
Stanley, P. (2017). *Charles Bean: Man, myth, legacy*. Sydney: UNSW Press.
Stockings, C. (2007). *The torch and the sword: A history of the army cadet movement in Australia*. Sydney: UNSW Press.
Stockings, C. (2012). *Anzac's dirty dozen: 12 myths of Australian military history*. Sydney: UNSW Press.
Strange, C. (2010). *The Personality of Environmental Prediction: Griffith Taylor as' Latter-day Prophet' Historical Records of Australian Science, 21*, 133–148.
Strange, C. (2012). Reconsidering the "tragic" Scott expedition: Cheerful masculine home-making in Antarctica, 1910–1913. *Journal of Social History, 46*, 66–88.
Strange, C., & Bashford, A. (2008). *Griffith Taylor: Visionary environmentalist explorer*. Canberra: National Library Australia.
Taylor, T. G. (1946). *Our evolving civilization: An introduction to Geopacifics, geographical aspects of the path toward world peace*. University of Toronto Press.
Taylor, T. G. (1951). *Geography in the Twentieth century: A study of growth, fields, techniques, aims, and trends*. New York: Philosophical Library.
Taylor, G. (1963). Geographers and world peace a plea for Geopacifics. *Geographical Research, 1*, 3–17.
Thom, B. (2017). *US Office of Naval research and the Australian coast vol 2017*. Sydney: Australian Coastal Society.
Thomson, D. F., & Peterson, N. (1983). *Donald Thomson in Arnhem land*. Compiled and introduced by Nicolas Peterson. Melbourne: Curry O'Neil.
Ubayasiri, K. (2015). The Anzac myth and the shaping of contemporary Australian war reportage media. *War & Conflict, 8*, 213–228.
Walker, B., & Salt, D. (2012). *Resilience thinking: Sustaining ecosystems and people in a changing world*. Washington: Island Press.
White, H. (2012). *The China choice: Why America should share power*. Oxford: Black.
Woods, M. (2016). *Where are our boys: How Newsmaps won the Great War*. Canberra: National Library of Australia.
Woodward, R. (2016). Military geography. In *International encyclopedia of geography: People, the earth, environment and technology*. Wiley. https://doi.org/10.1002/9781118786352.wbieg0280.

Chapter 3
'A Country Girt by Sea': An Overview of Australian Maritime Exploration and Policy Drivers

Peter Kenshole

1 Overview of Maritime Exploration and Oceanography in Australia

There is a common misconception that credits Lieutenant (later Captain) James Cook with the discovery of Australia in 1770. However, over 160 years prior to Cook's voyages of discovery there were encounters with the Australian coast by Dutch navigators, and during the mid-eighteenth to early nineteenth centuries French governments sent a range of scientific expeditions to Australia and the Pacific. See for example Clode (2007), Mundle (2015); also Collingridge (1895) for speculation of discovery in the fifteenth century and Adams (1986, p. 84-85) for speculation on the Portuguese reaching Australia in 1518 or 1522 and mapping part of the east coast. Adams (1986, p85) also mentions circumnavigation of New Guinea by the Spanish and Dutch prior to Cook's 1768–71 voyage to the region.

Nevertheless, Indigenous Australians had been inhabiting the country for over 50,000 years, and visits to Australia by Macassan traders prior to European settlement had also been recorded in Indigenous rock-art and in the journal of Matthew Flinders. Prior to the European influence there is also evidence of trade between Macassan fishers travelling the 'Malay road' and Indigenous inhabitants dating back around 400 years. The Indigenous relationship with the marine environment, while acknowledged and important, is excluded from the discussion in this paper.

Spanish trading ships also entered the region on passage to trade centres to the north, encountering the Australian coast when blown off course or when miscalculating the easterly advance along the trade winds and in search of trade routes. The earliest recorded European encounter with Australia was by Willem Janszoon in

P. Kenshole (✉)
Commander, Royal Australian Navy, Navy Capability Division, Department of Defence
Russell Offices, Canberra, Australia
e-mail: pete.kenshole@defence.gov.au

S. Pearson et al. (eds.), *Australian Contributions to Strategic and Military Geography*, Advances in Military Geosciences,
https://doi.org/10.1007/978-3-319-73408-8_3

1606 onboard the Dutch vessel *Duyfken (Little Dove)*[1] when he sighted the coast near Weipa in the Gulf of Carpentaria[2] and set foot ashore at the Pennefather River, 30 km north of Weipa. They did however believe that they were on land that formed part of New Guinea as the separation between New Guinea and Australia had not yet been discovered (Mundle 2015). The first known Englishman to set foot on Australian soil was the adventurer and pirate William Dampier in 1688.

The early visits to the coast of Australia can be largely attributed to the European spice trade with the East Indies (now Indonesia) and were incidental in their nature rather than voyages of discovery. Although many countries established trading with the region, the most dominant economically and in presence was the Dutch East India Company (VOC), which explored the region in search of new trading partners and opportunities. The earliest charting was triggered by a desire to discover easterly passages of trade through the archipelago from the VOC headquarters in the trading port of Batavia (now Jakarta). Janszoon's 1606 voyage for the VOC passed along the western coast of New Guinea and proceeded south-east into the Gulf of Carpentaria but failed in its secondary task to identify any passage eastwards into the Pacific Ocean. Later that same year, the Spanish captain Luís Vaz de Torres proceeded through the passage later named after him as Torres Strait by Alexander Dalrymple (the first hydrographer of the British Admiralty). Torres sailed along the southern coast of New Guinea but because the Spanish suppressed information about their exploration in the region, it was 156 years before the English and Dutch suspected the presence of the strait.

As trade with Batavia increased, there were pressures upon ship captains to find more efficient trade routes to reduce the amount of time taken to transport their valuable cargoes of spices and other goods. A new route was pioneered in 1610 by Henderick Brouwer in the *Rode Leeuw (Red Lion)* that halved the usual 12 month-long voyages from Europe to India and the Spice Islands. The unfortunate by-product of this passage via the Brouwer route at higher latitudes was that mis-timing the turn to the north resulted in ships encountering the treacherous western Australian coast. Over the next 120 years, this coast claimed the lives of hundreds of Dutch sailors and millions of guilders in VOC treasures. Navigation in the region only eased with the invention of the marine chronometer by John Harrison in 1737 which enabled the calculation of longitude.

Amongst the most notable of these encounters were by Dirk Hartog in 1616, Frederick de Houtman in 1619, Abel Tasman in 1644, William de Vlamingh in 1696, and William Dampier in 1699. All of these encounters were on the exposed western coast of the land and the general finding was that the land was inhospitable and barren. Hartog in the *Eendracht* made landfall on the west coast of Australia at Cape Inscription, Dirk Hartog Island, Shark Bay on 25 October 1616. Hartog nailed a pewter dinner plate to a post above the cliff at the bay with an inscription recording

[1]A full-scale replica of the *Duyfken* was constructed in 1999 in Western Australia. See https://www.mra.wa.gov.au/see-and-do/elizabeth-quay/attractions/duyfken

[2]The most prominent feature of this part of the coastline was later named Duyfken Point by Matthew Flinders.

their landing on this uncharted landmass. The plate[3] represents the oldest known evidence of European landing in Australia.

It was not until the 1768–1771 voyage of the *Endeavour* captained by Lt James Cook to observe the passage of Venus at Tahiti in 1770 that the British made a concerted effort to seek out the eastern coast of Australia in a true voyage of discovery. Ten years in the merchant Navy had provided Cook with an aptitude for navigation which was further honed through military service and conflicts with France. While the primary purpose of this voyage was to observe the transit of Venus, a second goal was to record natural history. It is noteworthy that the scientific element of the voyage was privately funded by Joseph Banks, and included Banks as expedition botanist along with the Swedish naturalist Daniel Solander and scientific illustrator Sydney Parkinson. A third but secret goal of the mission was to search for the 'Great South Land' and to take possession of the country 'in the name of the King of Great Britain.' As part of the exploratory component of the expedition, Cook was to verify the existence of Torres Strait.

After travelling to Tahiti and then New Zealand, the *Endeavour* first neared the Australian coast on 19 April 1770 at Point Hicks (named after the crewman who sighted the coast) in East Gippsland. The *Endeavour* proceeded northward along the coast seeking a suitable landing point eventually anchoring on 29 April at what he later named Botany Bay because of the great number of plants collected by Banks and Solander. The east coast proved to be a vastly different landscape to that of the west coast. A number of large bays that provided safe anchorage were located as well as sources of fresh water. Further north, a continuous reef was encountered dividing the deep oceanic waters and the protected shallow coastal zone. During the voyage, Cook created the first accurate charts of the east coast of Australia. He delineated the outline of New Zealand, and also confirmed the existence of Torres Strait. He claimed the eastern coast of Australia from latitude 38 degrees South to Possession Island, at the northern tip of Queensland in the Torres Strait, in the name of His Majesty King George III on 22 August 1770, together with all the bays, harbours, rivers and islands along that coast.

The British established a colony in New South Wales with the arrival of the First Fleet at Port Jackson in January 1788. The initial focus was on building self-sufficiency, which meant that coastal exploration was constrained at that time to the waterways surrounding the colony. It was not until 1795 when George Bass, Matthew Flinders and William Martin ventured south in a small boat called *Tom Thumb* that greater exploration of the coast was made. 1798–99 saw Bass and Flinders circumnavigate Tasmania, thereby proving it to be an island and opening the way for expansion of the colony beyond New South Wales. Matthew Flinders

[3] The Hartog plate was taken by De Vlamingh in 1697 back to Holland but he replaced it *in situ* with another that recorded both landings. The Dirk Hartog plate is now held by the Rijksmuseum, Amsterdam, although it has been loaned to the Australian National Maritime Museum, Sydney, for public display between 5 May and 29 October 2017, followed by the WA Maritime Museum, Fremantle, for six months from 31st October 2017. De Vlamingh's plate and a replica of the Hartog plate are held by the WA Maritime Museum, while a monument at Cape Inscription was installed in 1997 to commemorate the Dutch landings.

followed this voyage with the first circumnavigation and full charting of the Australian continent in HMS *Investigator*. Other notable surveys and charting were undertaken of the Australian coastline, Great Barrier Reef and Coral Sea by Jeffreys and King (1815–1822), by the *Beagle* (1837–1844), the *Rattlesnake* (1847–1850), and the *Herald* (1853–1860); see for example, Bowen and Bowen (2002). Charting in the nineteenth century thus reflected the growing size of the colony with a focus on the discovery and establishment of ports and safe anchorages and transport routes.

Following the work of these and other explorers, the British Admiralty established a Chart and Chronometer Depot in Sydney in 1897. The purpose of this depot was to supplement the survey activities of Royal Navy ships in Australian waters and to coordinate the charting of the coastal regions. In 1913, following Federation, the depot was taken over by the Australian Government and was renamed the RAN (Royal Australian Navy) Hydrographic Depot.

After the World War One (WW1), the focus of the British Admiralty was firmly on Europe and surveys of Australia were reduced to activities undertaken by a single vessel. This forced the Australian Government to create its own hydrographic surveying service which was considered best placed to hold regional naval responsibility. Consequently, the RAN Hydrographic Service was established on 1 October 1920 and assumed responsibility for regional hydrographic surveys. Once this initial work was completed in 1929, there was a short hiatus in survey operations (military or civilian) in Australian waters until 1933 when HMAS *Moresby* was recommissioned for survey work.

Surveying operations remained at a low level until WW2, when it became evident that charting of the South West Pacific was desperately needed as most of the charts still represented the needs of navigation in the days of sail. Several auxiliary ships, later supplemented by a number of Bathurst class corvettes, were modified into survey vessels. The tasks that these platforms concentrated on were focussed on the waters around the East Indies and New Guinea. In 1942, the RAN assumed responsibility for publication of charts, and in 1946, the Australian Commonwealth Naval Board was given surveying and charting authority for Australian waters and made responsible for supporting Allied operations in the South West Pacific Area.

In addition to updating navigational charts, RAN survey ships were also used to inspect and clear sites for amphibious landings. This is closely akin to the modern concept of Rapid Environmental Assessment which is the focus of the tactical level military survey activities. By the end of WW2, sixteen vessels were dedicated to survey activities. Following the war, the Australian Cabinet decided that the RAN would remain in control of all hydrographic operations in both Australian waters and areas of Australian interest.

Compared to the work undertaken to provide safety of shipping, dedicated research relating to the understanding of the physical properties of the ocean did not commence until much later. The first significant body of work on this aspect was undertaken by HMS *Challenger* through a global oceanographic voyage sponsored by the Royal Society of London. The voyage commenced in 1872 and lasted

1250 days, throughout which 713 days were spent at sea. The majority of 1874 was spent in the Pacific region and the Australasian region. The *Challenger* voyage is recognised as opening the way for modern oceanography.

In 1896, the American scientist Alexander Agassiz commenced the first of several expeditions to the Pacific Region. The focus of his work was to understand the formation of coral reefs predominant in the region.

The first dedicated Australian platform to contribute to oceanographic research was HMAS *Diamantina*. Recommissioned into service as an oceanographic survey ship in 1959, this River Class frigate worked in the Indian Ocean and waters around Christmas Island. Her work included discovery of the deepest known part of the Indian Ocean in 1960, named Diamantina Deep. *Diamantina* continued in service through to 1980, contributing greatly to the understanding of the Indian Ocean region.

The capabilities of *Diamantina* were replaced in 1980 by HMAS *Cook*. *Cook* was the first specifically designed oceanographic vessel to enter into service in Australia. She was also one of the first vessels in the world to be fitted with a deep water multi-beam system for deep ocean exploration. The SeaBeam system fitted to *Cook* enabled deep ocean observations with the vessel concentrating on building an understanding of the Western Pacific and Indian Ocean environments.

In 1985, ORV *Frankin* was commissioned as a purpose built oceanographic research vessel for the Australian Marine National Facility of the Commonwealth Scientific and Industrial Research Organisation (CSIRO). This began a period of military oceanography began a period of collaboration with government agencies and allied partners. HMAS *Cook* was decommissioned in 1990, ending dedicated military oceanographic surveys. *Frankin* was eventually replaced by RV *Southern Surveyor* and subsequently RV *Investigator*. While these proved to be effective research platforms, the foundation data created through this research was not tailored to meet military purposes nor was it targeted to areas of strategic interest for military operations. Reliance on allied data sources contributed to knowledge gaps of the operating environment.

2 The Importance of Regional Maritime Trade and Security

'A Country girt by Sea' is most apt to describe Australia and its engagement with the maritime environment. As an island nation with an economy derived predominantly from primary production, Australia is heavily reliant on seaborne trade. In 2014 the Australian Department of Infrastructure and Regional Development reported that over 99% of imports and exports (measured by volume) pass through the nation's ports. This equates to 1.62 billion tonnes of materials and cargo moving internationally plus an additional 51 million tonnes of domestic coastal trade.

Australia's broad continental shelf region to the northwest is a resource-rich area. During the resource boom of the early 2000s, the north-western region

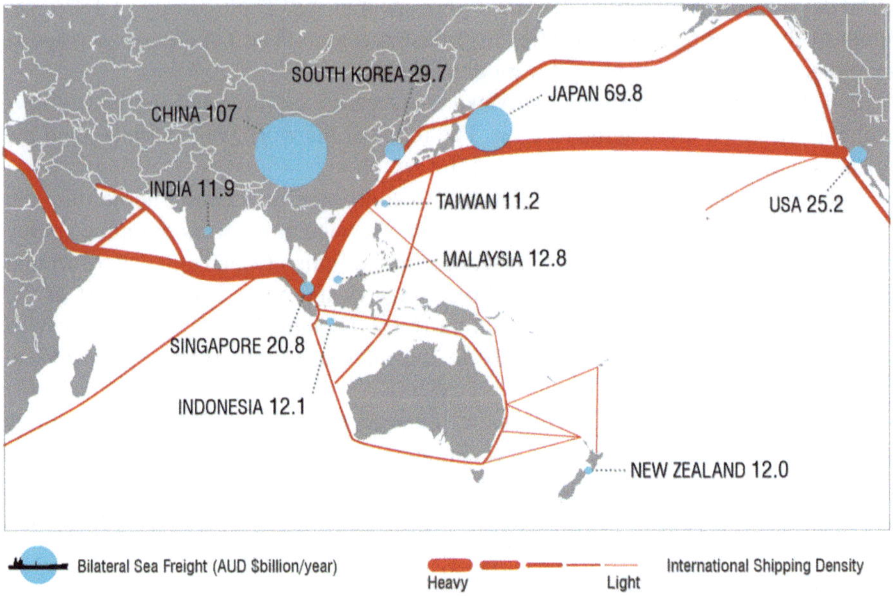

Fig. 3.1 The economic importance and density of sea freight between Australia and her top ten trading partners (Source: Department of Defence 2016)

produced a significant world market share of iron ore (21%), alumina (15%), nickel (12%), natural gas (9%) and diamonds (8%). The majority of these resources were transported to trade partners via a few important maritime routes (Fig. 3.1). Consequently, both the transportation of these resources and the long-term security of such resource-rich areas are of security concern. Ensuring a safe and secure environment for maritime trade is therefore one of the key drivers shaping Australia's modern strategic Defence capabilities.

This reliance on the maritime environment for trade has two important implications for Australian Defence priorities. The first implication is that a safe environment is needed within which trade can take place. Secondly, the trade routes need to be secure. Both of these factors necessitate a strong regulatory framework.

Maritime regulation is coordinated through the Australian Maritime Safety Authority (AMSA). The remit of AMSA includes providing the infrastructure to support safety of navigation in Australian waters, providing a national search and rescue service to both the maritime and aviation sectors, preventing and combating ship-sourced pollution in the marine environment, and promoting maritime safety and protection of the marine environment. These responsibilities are underpinned by Australia's commitment to the United Nations Convention on the Law of the Sea (UNCLOS) and the Safety of Life at Sea (SOLAS) Convention.

To achieve a safe trading, navigational and maritime operational environment requires a good understanding of the maritime environment. This includes both the littoral (coastal) and oceanic components of the maritime environment. In Australia, hydrographic services are provided by the Australian Hydrographic Service (http://www.hydro.gov.au/), which is part of the Australian Department of Defence. The primary focus of the Hydrographic Service is to provide and distribute products such as nautical charts to facilitate safe navigation by marine vessels. Operational surveying support and maritime military geographic information are also provided to assist with Australian Defence Force operations and exercises. The Australian maritime area of interest is depicted by the Search and Rescue (SAR) region (Figure 3.2), the region closely aligns with the area of charting responsibility. The Australian SAR region covers nearly 53 million square kilometres, which is one-tenth of the world's surface. The region encompasses all Australian territories and extends south to the waters of the Australian Antarctic territorial claim. The Australian SAR region has maritime boundaries on all sides that border the SAR regions of ten other countries.

In contrast to the SAR region *per se*, the area within which Australia provides hydrographic services closely echoes the SAR region but extends further north to encompass the waters of Papua New Guinea and the Solomon Islands. An agreement exists for Australia to produce charts on behalf of these two countries.

Not just Australia, but many countries in the surrounding region are also heavily reliant on maritime trade and navigational safety and security. For example, the developing nations of the South Pacific, and island nations of Papua New Guinea, Timor-Leste and Indonesia, all rely heavily on maritime trade. Furthermore, these countries also experience considerable challenges such as natural disasters, and pressures on economic development through high population growth and environmental degradation. Australian military capability is used to assist these countries with such challenges and to enhance regional stability. The significant investment that the country has made in Navy assets recently, including an amphibious capability and the desired increase in capacity and capability of the future submarines, reflects the strategic importance placed by the Australian Government on a secure maritime environment.

The 2016 Australian Defence White Paper identified that Australia's major trading partners include China, Japan and the United States. Trade routes to these countries pass through significant choke points such as Torres Strait, the Great Barrier Reef, and the Indonesian Archipelago to the north, and Papua New Guinea to the northeast. These choke points represent narrow areas through which shipping is naturally channelled, as indicated on Fig. 3.3 by areas of high density of ship reported positions (i.e. where there is a heavy convergence of trajectories). The figure shows the density of Automated Identification System (AIS) reports within the Australian coastal region. These indicate the locations of ships on a six-hour frequency mapped over a period of 12 months. As the majority of trade traffic to the north passes through these choke points ensuring these trade routes are secure is of vital importance to the Australian Defence Strategic Objectives discussed later.

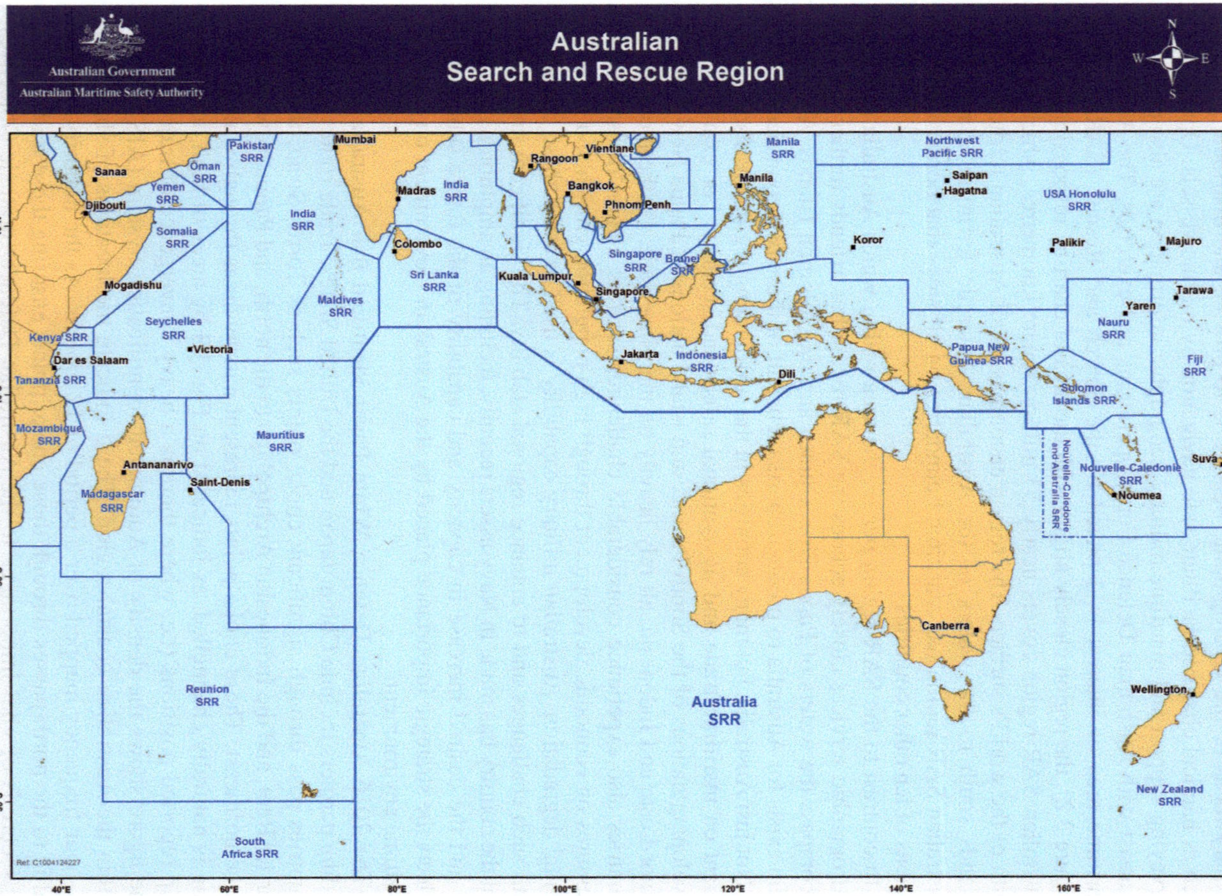

Fig. 3.2 Search and Rescue (SAR) regions of the south-western Pacific Ocean, Indian Ocean and adjacent Southern Ocean (Source: AMSA)

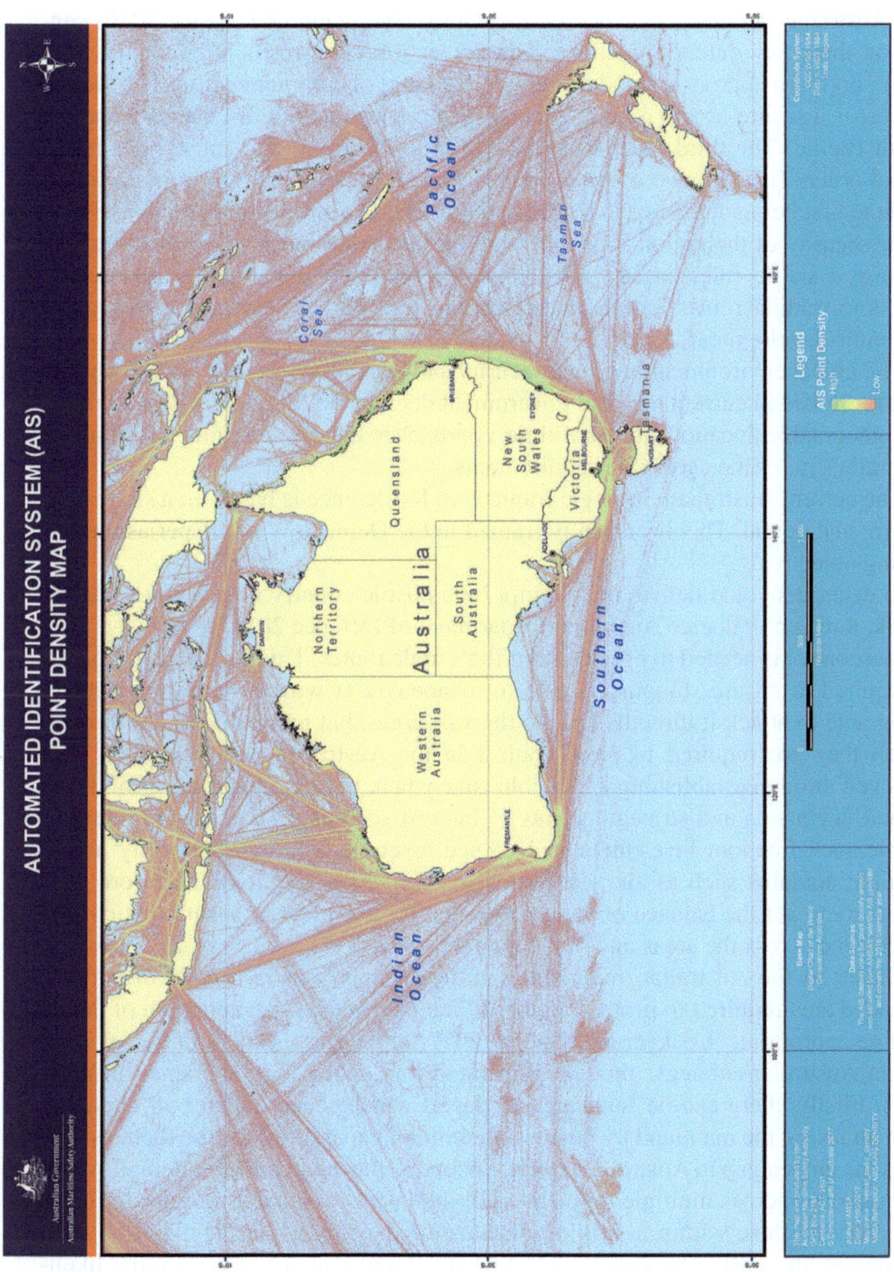

Fig. 3.3 Density of Automated Identification System (AIS) reports within the Australian coastal region based on ship locations at a six-hour frequency over 12 months (Source: AMSA 2017)

3 Strategic Importance of the Marine Environment

The Australian Strategic Defence Framework (Department of Defence 2016) centres on three key defence interests. The first is a secure, resilient Australia, with secure northern approaches and proximate sea lanes of communication. The second interest is a secure nearer region, encompassing maritime South East Asia and the South Pacific. The third interest is a stable Indo-Pacific region and a rules-based global order. The first two of these interests are predominately maritime in focus, with the first being defensive in nature and the second outward-oriented in assisting in the security of the region. The third interest also contributes to regional security because a stable rules-based global order facilitates regional safety and secure access to trading routes, communications and transport necessary for Australia's economic development.

The global environment, regional security and Australia's security and prosperity are thus linked in current national government defence policy and decision making. Our geographic location and maritime approaches clearly continue to influence Australia's maritime capability requirements.

The current Australian Strategic framework for defence is based on a strategy of control and denial. This is currently framed in the Defence White Paper as the following objective:

to deter, deny and defeat any attempt by a hostile country or non-state actor to attack, threaten or coerce Australia (Department of Defence 2016, p. 17).

Sea control is needed to protect Australia's trade routes. This command of the sea is required to ensure shipping is able to manoeuvre at will without an adversary being able to attack it directly. Due to the vast areas, but narrow passages through which they are required to pass, control in the Australian environment is best achieved through establishing a movable zone where military control is established around key assets and strategic points of interest such as the key trade routes. To enable such freedom investment in Defence assets to achieve superiority across multiple domains such as air, sea and cyber domains is required. Therefore, it is necessary to get the balance of capabilities and force correct to ensure that control is achieved across the areas and in the manner desired.

Closer to the Australian mainland a different approach is needed due to the extended area required to protect. In the littoral environment an approach of denial is more appropriate. Lockyer and Cohen (2017) in their assessment of denial strategy in Australian strategic thought illustrated that denial can be framed in three ways. Firstly, there can be 'anti access' denial which seeks to deter an opponent from attacking the mainland by possessing military capabilities to stop them as they cross the air-sea gap to Australia's north. Secondly, there can be 'area-denial' which also operates across multiple domains with the intent to make it too unsafe for an adversary to close within a striking distance of the Australian mainland. A third strategy is 'dissuasion by denial' which seeks to deter an adversary by the likelihood of incurring greater costs than the perceived potential political benefits if they were to successfully access areas of national responsibility.

The first two of these three denial approaches in particular drive a military capability response. The 'anti-access' strategy can be best achieved by maintaining and demonstrating military power that projects a superior capability to engage and defeat an opponent's forces, and thereby maintains sea and air control over the area of interest. However, given the Australian geographical and socio-political context (e.g. extensive border area, large SAR area, few large population centres in the north, relatively low national population size), this approach of regional control is the most logistically and economically demanding, and therefore difficult to maintain.

'Area denial' is an evolution of the concept of 'sea denial' and implies a more passive posture than anti-access. Area denial refers to obstructing an opponent from being able to safely manoeuvre within a defined space which extends to sea, land, air, space and more recently cyber domains. The advantages of adopting a posture of denial rather than control rely on the correct balance of military capabilities. A modern submarine force lends itself well to area denial as these vessels can exploit concealment, evasion, surprise and ambush tactics with the aim of denying control to an opponent.

The current Australian Defence White Paper, which places a stronger emphasis on submarines, strike aircraft and agile surface maritime platforms to achieve strategic defence objectives lends itself to a stance of area denial in the littoral environment while maintaining the ability to provide an umbrella of sea control further offshore.

Nevertheless, to exploit the full potential of maritime capabilities, superiority can be enhanced through advanced technology and a greater understanding of the maritime environments within which the technology will be employed. These aspects and the need for their further development are discussed in the next section.

4 Enhancing Australian Maritime Safety and Security

The Australian focus of surveying and charting concentrates on 'navigational critical areas' close to the coast. The relatively high density of maritime traffic (Figure 3.1), along with complex geographic and tidal conditions (e.g. Torres Strait), make these areas a challenge for all shipping activity. Yet, even today, only about 66% of the waters within the 200-metre bathymetric contour have been adequately surveyed.

In parallel with the ongoing navigational challenges, there has been a massive increase over approximately the last 20 years in average commercial container ship size and a concordant increase in transit speeds (average now between 20 and 25 knots). The driving factor in sea transportation is cost and therefore a desire to find more efficient and cost-effective vessel designs. These last two decades have seen the reliance on the canal system (including a widening project for both the Suez and Panama canals) reduced due to greater efficiency in propulsion systems. This has resulted in a rapid increase in vessel sizes. As a consequence, there is a need for improved understanding of the limitations and challenges of the maritime environment for modern shipping vessels and their activity.

Some Australian coastal areas are particularly vulnerable to damage from international shipping, unsustainable exploitation by commercial fisheries and changes to environmental conditions including water quality. These areas, including the Great Barrier Reef and Torres Strait, have been designated as 'Particularly Sensitive Sea Areas' (PSSAs). To provide a higher level of protection for these areas Australian regulations around PSSAs include requirements for ship reporting systems and compulsory pilotage (exempted are vessels with sovereign immunity, including defence and other government-owned vessels), although the latter requirement has been controversial under international regulations (Beckman 2007, Bateman & White 2009).

Tighter specifications in navigation products and complementary technological advancements are assisting to improve global shipping safety. An exemplar of maritime technological advancement has been the introduction of 'navikeel' clearance monitoring systems. Ten years ago, there was a two-metre clearance requirement under the keel of vessels passing through Torres Strait. However, in 2012, a dynamic under keel clearance monitoring system was introduced that enabled vessels to pass through the choke point with half the clearance previously required. Similarly, a notable improvement in navigation products has been the introduction of electronic charting systems. Electronic Navigation Charts are now an International Maritime Organisation mandated system. There has been a major undertaking by all hydrographic organisations to convert paper charts to electronic form. In Australia, the conversion activity which is now complete ran for six years. It is anticipated that next year the sale of paper navigational charts will cease. Importantly, electronic navigation drives the need for improved digital connection at sea to allow transfer of updates to take place while the vessel is underway and enable seamless integration of services such as Automatic Identification Systems (AIS) and weather services on demand.

In contrast to national maritime functions which focus on navigational safety and maintenance of sea trading routes, the national maritime military functions have a different emphasis. Firstly, the most common and most visible activity undertaken by the Australian military throughout the region is support to Humanitarian Aid and Disaster Relief operations. These activities accord with the second of the national strategic defence priorities of securing the stability of the region. To increase capability for aid and relief operations, there has been significant national investment in amphibious capabilities able to operate in affected areas. A second, sometimes more controversial capability has been demonstrated in supporting the third strategic priority of rules-based global order.

In the military context, there are very few Command and Control and mission systems that operate without geospatial data as a foundational layer to the system (Coorey 2017 this volume). Having comprehensive, validated geospatial information, including time sensitive data, requires detailed, accurate mapping and a common operating understanding across maritime, air and land domains. Knowledge of the physical attributes of the environment also provides tactical advantage. In particular, some physical properties of the ocean can be exploited for tactical advantage in submarine operations. The most common example is forecasting sonar performance and detection ranges. Water temperature profile, conductivity and salinity all contribute to the direction that sonar signals will travel. The type, hardness and gradient of the seabed also contribute to how sound is reflected and can dramatically change the range at which underwater noise can be heard. By knowing and mapping these attributes across the region, submarines can plan where and how to operate with the lowest probability of detection while maximising the performance of their own sensors and weapon systems. Recent commitment to further investment in submarines as part of Australian naval capability means that a refocus of national marine geospatial collection activities will be needed to ensure that adequate sourcing of relevant environmental data to support the national maritime policy priorities is achieved.

5 Conclusion

Being a relatively young nation, surrounded by sea and dependent on seaborne trade, the focus of much of Australia's historic maritime policy has been towards delivering navigational safety and trade security. The methodical and systematic surveys necessary to support navigational safety are time consuming and labour intensive. Although critical to the development and economic security of the nation, focusing solely on development of these products for commercial and civil outcomes detracts from the collection of environmental information for strategic defence purposes. The evolution of area denial as a key strategic approach to achieving defence priorities has emphasised the importance of enhancing environmental assets and information to support defence capability. There will be little advantage from an increase in the numbers and capabilities of strategic assets such as submarines (and amphibious craft) without understanding how to tactically traverse and exploit the environments within which they operate. Dedicated collection of littoral and maritime environmental information including bathymetry, nature of the seabed, reflectivity, water clarity, temperature, salinity and conductivity profiles will be required across the Australian maritime area of interest to ensure that the desired strategic intent is met. This will require a re-focus of the national marine geospatial collection activities with data unlikely to be able to be sourced through public databases or from allied nations but rather collected by dedicated maritime assets similar to the global surveys undertaken to support the introduction of the US and UK nuclear submarine fleets. Being 'girt by sea' has considerable advantages and constraints, and is thus an essential driver of Australian maritime defence policy and a fundamental Australian contribution to strategic and military geography.

Acknowledgment I would like to acknowledge Dr Jasmyn Lynch for additional research and editorial assistance in compilation of this chapter.

Disclaimer The views expressed are the author's and not necessarily those of the Department of Defence. The Commonwealth of Australia will not be legally responsible in contract, tort or otherwise, for any statement made in this paper.

References

Adams, B. (1986). *The Flowering of the Pacific*. Sydney: Collins.

Australian Maritime Safety Authority (2017). Automated Identification System (AIS) Point Density Map for 2016. Canberra, Australian Maritime Safety Authority.

Bateman, S., & White, M. (2009). Compulsory pilotage in the Torres Strait: Overcoming unacceptable risks to a sensitive marine environment. *Ocean Development & International Law, 40*, 184–203.

Beckman, R. C. (2007). PSSAs and transit passage-Australia's pilotage system in the Torres Strait challenges the IMO and UNCLOS. *Ocean Development & International Law, 38*, 325–357.

Bowen, J., & Bowen, M. (2002). *The Great Barrier Reef: History, science, heritage*. Cambridge, UK: Cambridge University Press.

Clode, D. (2007). *Voyages to the South Seas*. Melbourne: Miegunyah Press.

Collingridge, G. (1895). *The Discovery of Australia*. Hayes brothers, Sydney. Facsimile edition published 1983. Sydney: Golden Press.

Department of Defence. (2016). *Defence White Paper 2016*. Department of Defence, Canberra, Australia. Online at: http://www.defence.gov.au/WhitePaper/Docs/2016-Defence-White-Paper.pdf

Lockyer, A., & Cohen, D. (2017). Denial strategy in Australian strategic thought. *Australian Journal of International Affairs doi*. https://doi.org/10.1080/10357718.2017.127843.

Mundle, R. (2015). *Great south land*. Sydney: ABC Books.

Chapter 4
Reflections on Strategic Military Geography 2.0

Jane L. Holloway

> *"La politique des Etats est dans leur géographie"*
> *(The policies of all the powers are inherent in their geography)*
>
> *Napoleon Bonaparte (1804)*

1 Introduction

This chapter aims to introduce the reader to an iteration of strategic military geography dubbed SMG 2.0 and suggest ways its usefulness for addressing defence and security problems might be enhanced. The chapter provides a synopsis of how and why SMG2.0 developed within Australian Defence circles over the last 5 years, and briefly reviews long-standing geographic concepts including the terminology relevant to the later discussion. It goes on to describe SMG2.0 and the incorporation of recent developments, and complementary concepts including human cognition. The chapter concludes with some ideas about further development.

Starting from a realisation that biophysical global change was happening and would have implications for the Australian Defence Organisation (Defence), a small group of managers, researchers, and senior retired military personnel began to consider ways to incorporate geographical and systemic analyses into decision-making processes (Holloway 2012). Key drivers were the increasing likelihood of extreme and persistent climate and other Anthropogenic changes to Earth systems, the faith in the military to work effectively, impartially and ethically when all else fails (Barrie 2002) and mounting evidence that complex interactions were occurring between human, biophysical and cyber systems. This was accompanied by a growing awareness that current military systems may not be prepared for, or have the capability to respond to, these emergent challenges (see for example Dupont and Pearman 2006; Bergin and Townsend 2007; Templeman and Bergin 2008).

J. L. Holloway (✉)
Joint and Operations Division, Defence Science and Technology Group,
Department of Defence, Canberra, BC, 2610 ACT, Australia
e-mail: Jane.holloway@dst.defence.gov.au

S. Pearson et al. (eds.), *Australian Contributions to Strategic and Military Geography*, Advances in Military Geosciences,
https://doi.org/10.1007/978-3-319-73408-8_4

In addition, there was an emerging global and scientific consensus that planetary boundaries, which had provided a relatively safe and stable operating space for the human race for thousands of years, were being progressively and possibly irreversibly exceeded (Steffen et al. 2007). These complex, intertwined and novel challenges have obvious – and cascading - consequences for defence and security (Dupont and Pearman 2006; CNA 2007; Laksmana 2011; Boxall 2012). These implications at local, regional and global scales involve the United Nations, member governments and sub-national groupings (cities, local or provincial governments) who are increasingly mobilising to address national responses (Boxall 2012). Sometimes these include whole-of-government activities and military organisations, some of which – like the US Navy – have led identification and analysis of impacts (US Navy 2010). The Australian Defence Force is designed and indoctrinated to fight foreign (and mainly state-based) threats and so the question arises: Is it appropriately equipped and tasked to extend the concepts of defence and security to potentially catastrophic changes in (global and regional) biophysical geography? (Thomas 2012; Press et al. 2013).

2 Developing SMG 2.0

Military geography had waned as a subject of interest to academic schools, and the two component parts had separated and taken different pathways, academically and professionally. The cultural component had become part of Human Geography and was being rejuvenated as part of Critical Military Studies (Basham et al. 2015). Meanwhile, developments in the physical component of military geography had come to be dominated by geospatial information science and more lately the rise of cyber warfare and operations (Grant 2014). The development of SMG2.0 was partly an attempt to knit these two essential components back together, with a focus on the primacy of collective human agency.

Drawing together (or drawing upon) current and emerging themes in geography, such as Earth system science and cyber-geography, an analytical framework emerged which explicitly incorporated evidence and projections of biophysical change, human agency and influences and responses, complementing extant paradigms in preparedness planning and management. In doing so it became clear that different conceptualisations or ‘schools of thought’ reflected different ways of construing a situation or context.

A deliberative approach to how perspectives emerged was therefore included, so that such differences could be constructively exploited in developing responses to problems. Largely as a consequence, the approach included collaborating as widely as possible (within resource constraints) and managing a collective and sustained dialogue on pertinent issues. Cognitive level considerations also mattered because how issues are framed and conceived shapes the analysis about them and the responses to them. In the twenty-first century context of suites of inter-connected problems or challenges, greater complexity and diversity in thinking is needed to generate comprehensive, actionable responses that minimise unintended consequences (APSC 2007).

As the human operating environments become more complex and complicated, there arises a concomitant need for more complex, systemic cognitive tools and approaches to effectively address these suites of challenges (Bawden 2010; Boisot and McKelvey 2011). There is also the argument, originally propounded by Einstein (1946) and contemporaries, that problems arising now from earlier human activities are unlikely to be solved using the same kinds of conceptual approaches and framing that created them.

The core group was deliberately kept small–and therefore manageable. Personal relationships, shared purpose and commitment to Defence and a belief in the transformational power of deliberative experiential learning were fundamental to the coherence of the group. That is not to say that everyone agreed with everything all the time. What was consciously developed was a supportive atmosphere where ideas and issues – and underlying assumptions – could be discussed without any of the participants having to think about how to appropriately phrase and deliver any carefully thought through issues, proposals and solutions.

Another key determinant of success was that there were no preconceived or prepared plans, propositions or problems. Issues were identified and ideas elicited through interactive processes, such as workshops and free-form 'jam' sessions (think jazz jamming). The overarching methodological approach used to effect geographical enquiries was Systemic Action Research (Burns 2007), which is widely used in agriculture, community development and more recently, an earlier iteration has been adopted as a key methodology in Critical Military Studies (Basham et al. 2015).

The core group decided that it would be preferable to retain the words strategic and geography in the title of their approach. In social theory, the words in a title matter. The word order is partly related to where the problems and solutions are most likely to be found, but mostly to do with our value hierarchy (Ison 2010). As the SMG2.0 framework was designed to support a range of perspectives and methods, other Defence personnel drew upon the Cynefin Framework (Snowden and Boone 2007), particularly, and a range of geospatial, Cartesian and social science methods, so that as SMG2.0 evolved, the accompanying pool of intellectual tools grew also.

The initial aims of developing SMG2.0 as a holocentric (collective evolutionary learning) and geographical inquiry framework were to:

1. Put more 'geo' into Defence decision-making and geo-strategic thinking;
2. Insert physical geographical change into Defence decision making, including developments such as the advent of a new geological era (The Anthropocene), Earth System Science, climate risk and resilience;
3. Engender and support strategic, systemic analysis;
4. Facilitate collaborative and co-ordinated action (choreography).

Outcomes included: (1) a broad consensus that physical change is occurring and will affect Defence in direct and indirect ways, at the strategic, operational and tactical levels, (2) awareness-raising and (3) a way to characterise contentious or controversial issues in a neutral and discussable way. Examples of such issues include climate change and synthetic chemical pollutants as well as national and community responses to change impacts.

3 SMG 2.0 Underpinning Geographic Concepts and Terminology

Geography, as an integrative and applied science, is able to use the tools and understanding of human and biophysical sciences, and drawing on the knowledge of many disciplines, to solve and resolve problems that defy discipline-specific solutions. Integrating geography with complex adaptive systems and system-of-systems approaches provides an engaging, effective and appropriate way to deal with global changes that may affect Defence strategy and operations. Geography offers tools to know place and space, time and scale that can be used to identify and work with systems of interest, and used purposefully to achieve desirable outcomes.

A detailed definition devised by Australian geography institutions (Australian Academy of Science et al. 2010) alerts or reminds people of the scope and depth of geography enabled thought:

> *Geographical thought develops knowledge and understanding from three complementary concepts.*
>
> *The first is the concept of **place**. Geographers explore what places mean, how people shape places, and how places shape our lives. This brings many areas of geography together in a holistic approach to understanding the characteristics of, and relationships between localities, cities, regions, countries and continents.*
>
> *The second concept is **environment**. Geographers investigate biophysical environments encompassing terrestrial, marine and atmospheric systems. These investigations include the nature, dimensions and causes of environmental change; the reciprocal relationships between the environment and people; the resources biophysical systems provide and their sustainability.*
>
> *The third concept is **space**. Geographers examine how, why and with what effect diverse phenomena vary across the surface of the earth. Geographers understand space to be configured by the movement and organisation of people and materials as well as being a location for social and material action.*

In the twenty-first century, it is also about constructing and exploring entirely new environments such as human-machine interfaces, virtual reality, holographic simulations and orbital space via new kinds of places, such as the International Space Station.

Military Geography is a sub-discipline of geography.

> *"Military activities occur on landscapes with distinct physical and cultural character. Understanding how the spatial distribution of landscape elements affects the military operating environment at the tactical and operational scales, and how this applies to military concerns at the strategic scale, is the substance of military geography."* (Harmon et al. 2004)

Military geography is thus defined as, "the application of geospatial concepts, approaches and tools to military problems in war; military operations other than war, such as peacekeeping; and military activities during peace-time, such as disaster relief" (Galgano and Palka 2011).

The two prime aspects of military geography are the physical landscape and associated cultural characteristics. These are mapped, analysed and planned for at different spatial scales and across various timeframes, depending on the military activity being undertaken, e.g. the strategy being devised or operation being planned or conducted. The physical environment includes any feature or characteristic that may help or hinder military activity, whether it be at the tactical, operational or strategic level (Collins 1998). Cultural aspects of military geography have expanded over the years and become more prominent in operational and strategic activities, particularly after World War 2 and during the Cold War (Galgano 2017).

4 Strategic Military Geography 2.0

At the strategic level of contemporary military geography, the aim is to understand changes in the biophysical, human and, increasingly, cyber environments that are likely to have implications and consequences for defence and national security (Durrant and Holloway 2013). Climate change, for example, is adding to the complexity of strategic planning, preparedness and operations (see for example Dupont and Pearman 2006, Barnett and Adger 2007, Laksmana 2011, Holloway et al. 2014).

The recent strategic military geography iteration, dubbed SMG2.0, was developed in response to a seeming barrage of changes across multiple domains, and a perceived need for new or different ways of perceiving, understanding and responding to both the manifold changes and their interactive effects. The name reflects acknowledgement of historical concepts and activities in strategic military geography, and recent evolutions and expansions, not to mention the rise of the meme!

In pursuing these aims, SMG 2.0 was developed using approaches or techniques that support:

1. Thinking big picture; looking at the whole set of challenges and potential changes in the defence operating environment, to get a sense of the range of issues and how they interact and influence each other.
2. Considering the relationships between the issues as well as understanding specific issues.
3. Taking a participatory approach and engaging as broadly as possible – within Defence, across government and industry, and within the science and research communities.
4. Consciously using a multi-level learning process.

SMG 2.0 also draws upon philosophy and concepts in epistemology (the nature, origins and limits of human knowledge, e.g. how do we know what is "true"? Kvanvig 2003), axiology (the study of value, e.g. how do we determine what is "the right thing to do"? Rescher 2005) and ontology (the study of the nature and categories of existence, e.g. what does it mean to be systemic? Harvey 2006). The Hawkesbury heuristic, discussed later in this chapter, provides for deliberative development of systemic competencies, critical self and group reflective practice, and use of multi-level learning.

SMG2.0 can be useful for understanding where military forces sit within and across broader systems, and to focus thinking on how to integrate these different aspects to gain a richer and more comprehensive appreciation of the Defence operating context, in the short, medium and longer term. Thus SMG2.0 helps to both normalise defence and national security in the modern geography context and to securitise environmental change in a military context, where this may be useful (Holloway et al. 2014).

Importantly, SMG2.0 also explores the insights that may be gained from different views about the conceptual framework and drivers for military geography. For example, Griffith Taylor (1963), one of Australia's greatest geographers, espoused an alternative to 'geographies of war', advocating that geography and military students should instead be taught 'geopacifics' – the geographies of peace and how to attain and retain it through the practices of geography and related disciplines. This concept still reverberates and resonates today, in Defence and broader society. Within the Defence community, geopacifics is seen as a constructive and useable concept, particularly compared to some of the more strident critiques of militarism (e.g. Woodward 2004).

Geopacifics and similar constructs deliberately reframe and change the focus of military thinking from the task of defeating an opposing force, to a more co-operative, consensus-seeking mindset. Such reframing can, for example, engender new insights and alternative pathways to resolution of problems that appear to only be solvable by combat operations. The United Nations initiative on Women, Peace and Security implicitly reflects a reframing of conflict that includes short to long term impacts on non-combatants, and therefore a nation's capacity to recover from the aftermath of war (UNSCR 2000).

At the strategic level of planning and preparation, this idea of framing is arguably fundamental and deserving of greater explicit attention. By deliberatively including framing in developing strategic responses, a diversity of perspectives and world-views (or more comprehensively *Weltanschauungen*; Smith 1956) can be drawn upon to find underlying commonalities and constructively exploit differences. Such conscious elicitation and use of epistemological capabilities can diminish risks such as 'group think', assumed like-mindedness or difference, responses that are too simplistic or complicated, gaps in a suite of responses, and insufficient creativity or innovation due to the homogenising and/or reductionist effects of many decision-making processes (Bateson 1972).

4.1 SMG 2.0 Physical Geography Aspects:

While using current physical geography principles, concepts, research methods and tools, including geospatial technologies and analytical techniques, SMG2.0 also specifically incorporates Earth system science, particularly the planetary boundaries construct. The focus is on geospatial concepts and approaches to describe the Earth system and significant interactions that may be relevant to long term defence and security issues, particularly the future operating context and likely tasking.

The global climate and ecology are changing, largely but not exclusively driven by the cumulative effects of human activity (Rockström et al. 2009). Governments and societies are beginning to respond to these changes, including particularly climate change. Some impacts of these global changes, especially land use change, biodiversity loss and climate change, are already observable and there is broad scientific consensus that further change will occur (IPCC 2013). These changes will affect Australia across all sectors and in all ecosystems (Cleugh et al. 2011). These observations led to a set of propositions, which were used in framing the biophysical component of SMG2.0:

1. Human activity is now so pervasive and diverse, it almost covers the entire planet (Crutzen 2006).
2. The cumulative output of human endeavour is measurably having effects on the biological, chemical and geological systems of the Earth (Rockström et al. 2009).
3. Human-kind has become so numerous and powerful that it now rivals the great forces of nature (Steffen et al. 2007).
4. It has become obvious that the Earth's systems are responding (Reid et al. 2010).

This more comprehensive understanding of our world has come about through a keener appreciation of the Earth as a complex adaptive system, and the collective agency of humans within this system. These developments have also been primary drivers of the advent of a new discipline called Earth System Science (ESS) (Ehlers and Krafft 2006), which focusses on the Earth as a complex biophysical system.

In particular, ESS considers interactions between the Earth's "spheres"— atmosphere, hydrosphere, cryosphere, geosphere, pedosphere, biosphere and the magnetosphere—as well as the impact on these sub-systems of human activities (Crutzen 2006) – what could be termed the Human Piosphere! Like the broader disciplines of systems science and physical geography, ESS takes a holistic view of the dynamic interactions between the Earth's spheres, the variety of component sub-systems, their spatial and temporal developments, and to what extent they are stable at any particular point. Examination of such dealings are integral to geography – from the local or site-specific (endemic), through the regional to the global scale (systemic).

A key part of ESS is the concept of a safe operating space for humanity, expressed through nine planetary boundaries and the development of criteria for each boundary. These boundaries reflect biophysical processes of the Earth system that determine the self-regulating capacity of the planet, and which will ultimately limit continuous expansion of human populations and activity, though not necessarily prosperity. As shown in Fig. 4.1 so far, nine parameters of the Earth system have been identified, for which measurable boundaries or thresholds are considered imperative. These are: climate change; biodiversity loss; excess nitrogen and phosphorus production, both of which pollute our soils and waters; stratospheric ozone depletion; ocean acidification; global consumption of freshwater; change in land use for agriculture; air pollution; and chemical pollution (Rockström et al. 2009). Expert assessment of the current status of these boundaries is that several boundaries have already been exceeded, and four more are at risk. Two boundaries and one sub-boundary currently have no assigned thresholds, due to lack of data and/or difficulties measuring change. Staying within these boundaries or thresholds is crucial.

The theory is that exceeding one or more of these thresholds may trigger non-linear changes in the functioning of the Earth system, thereby challenging social-ecological resilience at local and regional to global scales. The current thresholds reflect a conservative, risk-averse approach to quantifying the planetary boundaries and take into account the uncertainties surrounding exact quantification. Notwithstanding this, the planetary boundaries construct is being used for research and assessments of global change and the types of responses required (see for example Raworth 2012).

This planetary-boundaries approach links strongly to geographical enquiry and provides a useful basis for identifying and understanding consequential changes in the operating contexts for military forces and defence organisations.

4.2 SMG 2.0 Human Geography Aspects

Modern military geography includes demographics as well as less tangible aspects of populations such as morale, character, language, customs, values and beliefs (*Weltanschauung* collectively), which describes the fundamental cognitive orientation of an individual or society. This is important for understanding the structure and functioning of nations or groups of interest, particularly for example legal and governance arrangements, and willingness to defend, or try to impose on others, those beliefs and values. Cultural aspects also include human artefacts, principally the built environment including transport infrastructure, telecommunications, trading routes and transport logistics (Collins 1998).

Here also, SMG2.0 uses contemporary human geography themes, memes and philosophies, in line with the military geography sub-discipline where the broader sub-discipline of human geography has replaced cultural geography. Human geography also identifies and analyses social, technological and economic responses and includes diverse elements such as domestic and international regulation, economic instruments, societal (non-government) reactions and Government activity as well as broader human activities, such as changing patterns of settlement, human demography, impact of technology applications and regional/global relationships including

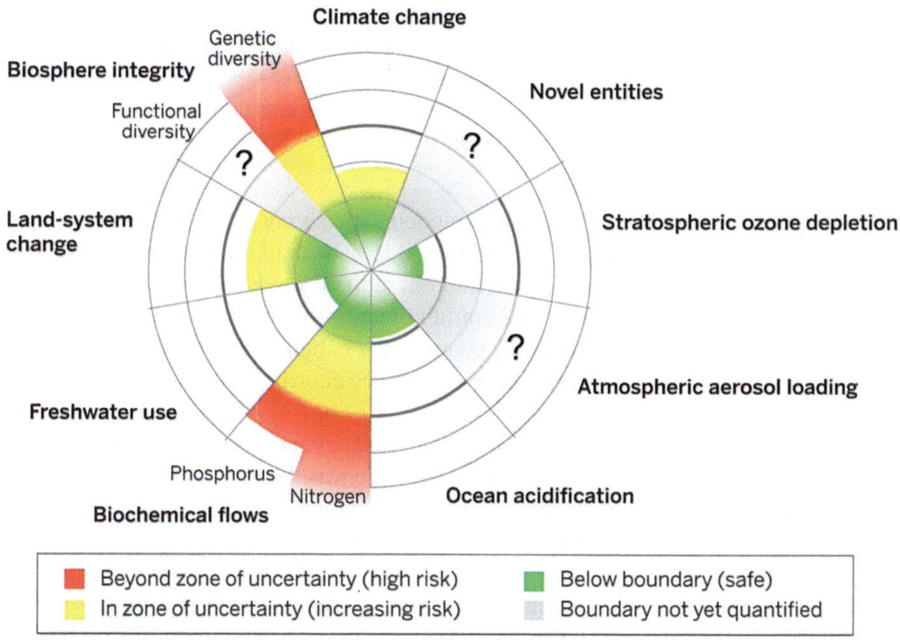

Fig. 4.1 The nine planetary boundaries; a summary of the evaluation (Steffen et al. 2015)

disasters and conflict. Particularly used in military geography context are cultural, political, regional and economic aspects (Collins 1998).

In a Defence context this also includes strategy, doctrine, force structure and disposition, the arts of peace, preparation, war and post-conflict reconstruction policies and practices. Thus, philosophical and conceptual geographies are also a focus in SMG2.0, that is the spatial and temporal nearness (sympathy, compatibility, oppositional or other connection) of concepts, theories and ways of understanding (epistemologies). The focus is on what is relevant to strategic analysis, particularly for preparedness and operations.

4.3 SMG 2.0 Cyber Geography Aspects

Technological developments globally over the last 20 years have engendered discussions about a new sub-discipline – Cyber Geography. See Kitchin (1998) and Warf (2001) for seminal examples. Using the definition of geography described earlier, cyber geography encompasses:

1. Spatiality, differently imagined but still within the broad definition of geography
2. Temporality, micro-temporal, speed of transmission, global traversement near-instantly (faster than humans can keep up with/manage) requires different

conceptualisation, strategies, management, collaborations and operational parameters. This could, for example, be exploited for humanitarian assistance and disaster relief (HADR), peacekeeping and warfare.
3. Human relationship with their environment; how humans deal with cyber terrain, how cyber operations or activities can influence the social-political-economic-epistemological topography.

The cyber-sphere appears to have exploded rather than expanded, and defence organisations are increasingly focussing on this new domain. Cyber-security and cyber-warfare are now major research and operational fields (Singer and Friedman 2014), in addition to information and communications technologies. Cyber geography has emerged over the past 15–20 years and it is being increasingly securitised through the development of Australia's first Cyber Security Strategy (Department of Prime Minister and Cabinet 2016).

In 2016–17, following well-reported global cyber security incidents, the Australian Signals Directorate, situated in Defence was directed by the Minister, "to use its offensive cyber capabilities to disrupt, degrade, deny and deter organised offshore cyber criminals" (Payne 2017). Use of this capability was previously limited to helping target, disrupt and defeat terrorist organisations. This new directive maintains consistency with Australia's international legal obligations, and is subject to stringent legal oversight. It also illustrates the 'boomerang effect' of capabilities deployed abroad by the military 'coming home' and crossing into civilian life even when they may be at odds with Australian values (Kilcullen 2016).

The number and variety of issues with defence implications continues to grow. There is increasing reliance on the internet for management and communication, for example of power grids, nuclear reactors and transport systems. There are potential cost savings by using virtual worlds to plan, train and exercise. What happens in the ether has impacts in the physical world. There are other linkages to the physical world, for example powering, location, development, trade and disposal of the hardware associated with the virtual world.

5 Enhancing SMG2.0: Integrating Framing

At this stage, with SMG 2.0 explained, some potential enhancements to the approach might be usefully discussed. The first improvement worth considering appears to be by integrating cognition. This was because as Defence issues become more complex, it is arguable that there is a concomitant need to develop systemic intellectual capabilities to address such complexity and inter-connections (Boisot and McKelvey 2011). In then considering various new intellectual capability alternatives, the group placed some emphasis on the observation (or plea) attributed to Albert Einstein (1946) that we cannot solve today's problems with the same type of thinking that created them. How we think about things, clearly shapes what we do about them and this makes cognition – our thinking processes – crucial (Bawden 2010).

In this, the key cognitive consideration in SMG 2.0 seems to relate to framing given this seriously impacts the coherence of any problem-solving responses Defence might adopt. The proposition advanced here is that these responses need to be developed within a cognitive frame of Earth as a tightly coupled socio-techno-ecological system of systems.

The concept 'socio-techno-ecological system' is an epistemological device, a way of knowing about our circumstances that, like the operation of metaphors, reveals and conceals. It reveals the dynamically coupled nature of humans to 'eco-systems' but at the same times conceals the idea that humans are 'part of nature' and that ecosystems are essentially constructs rather than entities in and of themselves (Ison et al. 2010). This proposed systems construct is holocentric (collective, evolutionary and epistemic 'human system' learning), but appreciative of the ecocentric (environmental influences), the technocentric (enabling or constraining technologies) and the egocentric (individual *Weltanschauung*) (Bawden 2010).

Consciously taking a systems approach as a frame can move people away from a linear cause-effect-response frame so they much better appreciate today's more complex situations where the obvious short-term response can be investigated and weighed against alternatives in different spheres of activity (or constituent systems) and over different timescales (Bawden and Packham 1993). Adopting a systems frame can also be a useful aid to understanding of the range of externalities associated with different courses of action.

In this vein, it seemed pertinent to the SMG 2.0 development group to consider not only what issues were chosen for analysis and how they were chosen, but also the ways in which the intellectual competencies underlying these activities could be themselves analysed and developed further. A heuristic approach to systemic development proved both compatible and useful.

6 Improving SMG2.0: Integrating Cognition

Underpinning the SMG2.0 approach are key components including the development of systemic competence, a holocentric (or evolutionary collective learning) orientation and conscious inclusion of multi-level learning. The Hawkesbury Heuristic exemplifies these key components. In this section the Hawkesbury Heuristic is briefly discussed along with the role that experiential learning and multi-level learning play in enhancing collaborative strategic analysis.

6.1 Enhancing SMG 2.0: Thinking using the Hawkesbury Heuristic

It is not the strongest of the species that survive, nor the most intelligent, but the one most responsive to change (Megginson 1963)

The Hawkesbury heuristic of systemic development and systemic action enquiry integrates systems sciences and learning approaches to address complex, connected issues in large social-ecological and/or organisational contexts. Some reasons to take a systemic approach to organizational management include the complexity of many issues, the need for cross "silo" collaboration, strategic uncertainty, operational change, a lack of clarity in roles, responsibilities and relationships (both between silos and across issues) and externally driven change (or system perturbation) (Bawden 2010).

In this heuristic, to effectively address a range of problems (including scientific, technological, economic and socio-political), the world is viewed as interacting sets of human activity systems. This notion recognises that humans are to a great degree capable of defining for themselves the purpose of their own behaviour and the systems of which they may be a component. That is, the human elements, individually and collectively, have a degree of autonomy to define the purpose of that system. So, an essential aspect of human activity systems is that they exist to do something. In order to discover what it is that these systems are meant to do, it is necessary to take account of the *Weltanschauungen* of the people involved in them (Bawden and Packham 1993). This aligns with the cultural assessments made in military geography, e.g. in determining a country's strategic interests or whether a population's defensive posture is pacific or belligerent (Collins 1998).

This does not, however, mean uncritically accepting the worldview of any or all of the people involved in a particular system. Most often, it means consciously articulating people's worldviews, and the assumptions that arise from them, as an implicit part of systemic inquiry and as a critical step in satisfactorily addressing problematic situations (Bawden and Packham 1993).

These distinctions and definitions are important because of the way different approaches can shape the problem definition, analysis and subsequent solution options or recommendations. How a problem is viewed initially – and what the roots of the problem are – is often a judgement call. In this context, judgement is defined as a combination of experience, knowledge, and learning preferences. The Hawkesbury heuristic not only accepts, but positively encourages a pluralistic philosophical stance.

"We are faced with a variety of ways of knowing, based on different philosophical positions; disputable propositions about the nature of knowing. Even in those situations where the 'facts of the matter' seem irrefutable, even when these facts can be seen to be manifestations of the most powerful of theories, describing the way nature is patterned, even where such theories are such strongly testable generalisations that they can be granted the status of 'laws', all seem to be eternally tentative." (Bawden and Packham 1993)

6.2 Enhancing SMG2.0: Embracing Learning Approaches

> *"Mental models are the images, assumptions and stories which we carry in our minds of ourselves, other people, institutions and every aspect of the world. Like a pane of glass framing and subtly distorting our vision, mental models determine what we see. Human beings cannot navigate through the complex environments of our world without these 'mental maps'; and all of these maps, by definition, are flawed in some way."* Senge (1994)

In the earlier discussions, the impact of the complicated and complex nature of the contemporary environment in making decision-making and problem-solving more problematic has been evident. Given this, the application of SMG 2.0 needs to be humble in the sense that the first solution devised may not be the most effective or efficient. It may be necessary to change this solution – or adopt a completely new one – as its success of otherwise becomes apparent. To do this, incorporating learning into the SMG 2.0 paradigm is fundamental: both experiential learning and multi-level learning relate to systemic development.

Learning here is "the process by which personal experience (actual and imagined) is transformed into knowledge (meaning) for action." (Bawden and Packham 1993). Experiential learning is seen as the best way to incorporate both theory and practice, and propositional and practical learning in an integrated and meaningful way. It is essential to development of systemic competence (Bawden 1997) and thus to applying SMG 2.0 in real-world circumstances.

To do this though, the idea of learning though needs some unpacking. People learn and change at three main levels, as proposed by Kitchener (1983) and further developed principally by Senge (1990), Bawden (1992) and Argyris (2006). Each level is a foundation for the next one but remains distinct from it. While people may function at the first level independently of the second and third, it does not work in reverse. The second level is dependent on the first, and the third works in conjunction with the first two levels.

At the first level – cognition or single loop learning – the learner does cognitive tasks such as reading, perceiving, memorising. This is the level of data gathering and knowledge acquisition, and is mainly concerned with actions. Argyris and Schön (1996) characterised this as the most common type of learning – new information and skills that change our practices, and enable better performance.

The second level – metacognition or double loop learning – as defined by Kitchener (1983) is concerned with learning about how one can learn. This includes how to carry out cognitive tasks, particular strategies that will support the task and how and when such strategies should be used, as well as examining the efficacy of any of these processes. Argyris and Schön (1996) propose that this entails reframing and taking a different perspective; considering our assumptions and examining our mental framework. It leads to insights about why a solution works, and can therefore be applied again with some confidence.

The third level – epistemic cognition or triple loop learning – addresses the nature of what can be known or learned. Kitchener (1983) contends that triple loop learning relates to the processes used to understand the limits of knowing, the certainty of knowing and the criteria for knowing. It also includes the strategies that can be used to understand and select from a range of solutions (or methods) to address different types of problems. Argyris and Schön (1996) view it as profound, or transformative, learning, i.e. affecting not merely our behaviour and way of thinking, but how we view ourselves and the world, and thus our very identity. This type of learning involves principles and context. It is fundamental to the Hawkesbury heuristic because it helps to enhance ways to comprehend and change our purpose, develop better understandings of how to respond to our evolving environments, and deepen our comprehension of why we choose to do things we do.

A quick synopsis of triple loop learning used in military circles (albeit not called that) is:

1. Are we doing things right? Here's what to do—procedures or rules. Also called TTPs (tools, techniques, practices) and reported with Measures of Performance (or Efficiency).
2. Are we doing the right things? Here's why this works—insights and patterns. Operational objectives/mission aims can be included here and reported with Measures of Effectiveness.
3. How do we know what's right? Here's why we want to be doing this. Principles, values, perspectives, societal expectations and so on are included here. This could be guided by doctrine and is perhaps reported with Measures of Strategic Intent.

The aim here is to both enhance and to share individual learning in order to develop 'organisational learning', and therefore support a holocentric (Bawden 1997), or evolutionary collective learning, approach to the development of systemic competencies, and of capabilities to act in an increasingly complex, complicated and rapidly evolving world. In responding to threats and opportunities, Bawden and others, e.g. Röling 2002, have demonstrated that in taking a holocentric approach, the emergent strategy will reflect the views and values across the group or community, as well as the creative tensions inherent in any negotiation process.

Managed well, this type of dialogue will engender more effective solutions that are better supported, more equitable, have fewer unintended consequences and which can minimise the chances of future conflict. It is particularly useful for diverse groups, where there are likely to be distinct differences of opinions, perspectives and worldviews. As a group or community becomes more practiced at such earnest, values-based dialogues, individual and collective competence increases, which may in turn further enhance the overall level of understandings and quality of responses. Arguably, in an increasingly complex world there is a need for concomitantly complex and collective appreciations of the suite of interactive changes that seem to be rapidly evolving, and their defence and security implications.

7 Reflections: Where to from here?

In a complicated, complex and tightly coupled world (e.g. Homer-Dixon et al. 2015), SMG 2.0 is about connections across space, time and communities. As Senge (1990) propounded, *"for the first time in history, humankind has the capacity to create far more information than anyone can absorb, to foster far greater interdependency than anyone can manage, and to accelerate change far faster than anyone's ability to keep pace."*

To remain operationally effective, in this more complex and connected world, it is useful for defence forces to consider not only new domains, doctrine and training, but also new ways of comprehending and analysing (Churchman 1971).

Geography remains fundamental in military affairs especially preparedness (readiness), strategy and operations. The SMG2.0 framework was developed as a means of thoughtfully engaging in collaborative exploration of relationships between issues of concern, such as global biophysical change, and military activities, through a geographical lens. Insights are derived from how geographies are arranged internally and in relation to each other, and how they provide a context for understanding and taking action on issue(s) of concern.

The definition of SMG2.0 is built on modern military geography, and the broader definition of geography, as briefly described earlier. The key inclusions in SMG2.0, in line with discussions in these larger communities, are:

1. The inclusion of orbital and Solar-system space in Physical Geography as well as emerging concepts of the Anthropocene, Earth System Science and planetary boundaries;
2. A change from Cultural to the broader sub-discipline of Human Geography, and drawing upon Critical Military Studies which has emerged as the revival of Military Geography after some decades of being largely ignored by Human Geographers; and
3. The recognition of cyber as a distinct geographical realm and emerging sub-discipline, and incorporating into this geospatial information sciences as well as remote sensing, 'big data', semi or fully autonomous systems information systems and virtual reality.

The emphasis remains on seeking to understand the connections, interactions and influences across and between changes in the three geographic domains at a holistic or whole system level, and at the same time, reflecting on how these processes are conducted and could be enhanced. This included adapting the processes, and even changing the issues under consideration, as inquiries proceeded. It was seen as a useful alternative means of defining a problem or set of connected problems, especially those subject to a great deal of uncertainty in how they might manifest.

Incorporating the Hawkesbury heuristic into SMG2.0 was done deliberately – to draw attention to the supportive or enabling theories, concepts, methods, tools and techniques, particularly in terms of individual and shared evolutionary learning, and systemic development.

Further development of SMG2.0, would benefit from a focus on: cyber geography structural integration of systemic and holocentric thinking and learning approaches, and incorporation of the Hawkesbury heuristic; application of SMG2.0 at different scales, and greater engagement between geographers, military geographers, regional geographers, systems scientists, learning researchers and Earth system scientists.

In the near to medium future, SMG2.0 could include: climate change impacts on occupational health, liveability and productivity in Defence establishments, emerging risk profiles of pandemics and complex emergencies and regional risk assessments elaborating improved insights on what constitutes Defence resilience, organisational design, adaptation and novel technology effects and fractal approaches.

References

APSC. (2007). *Tackling wicked problems: A public policy perspective*. Australian Government, Canberra, Australia. Online at: http://www.apsc.gov.au/publications-and-media/archive/publications-archive/tackling-wicked-problems.

Argyris, C. (2006). *Triple Loop Learning*. Online at Thorsten's Wiki, http://www.thorsten.org/wiki/index.php?title=Triple_Loop_Learning [downloaded 10 January 2011].

Argyris, C., & Schön, D. (1996). *Organizational learning II: Theory, method and practice*. Reading: Addison Wesley.

Australian Academy of Science - National Committee for Geography, Australian Geography Teachers' Association, Geographical Society of New South Wales, Institute of Australian Geographers, Royal Geographical Society of Queensland, and Royal Geographical Society of South Australia. (2010). *What is Geography? Nature and extent of geography*. Online at: https://www.iag.org.au/what-is-geography- [Accessesd 23 Oct 2017].

Barnett, J., & Adger, W. N. (2007). Climate change, human security and violent conflict. *Political Geography, 26*(6), 639–655. https://doi.org/10.1016/j.polgeo.2007.03.003.

Barrie, C. A. (2002). *The Australian approach to warfare*. Department of Defence, Canberra, Australia. Online at: http://www.defence.gov.au/publications/taatw.pdf [Accessed 27 Nov 2017].

Basham, V. M., Belkin, A., & Gifkins, J. (2015). What is critical military studies? *Journal of Critical Military Studies, 1*(1), 1–2. https://doi.org/10.1080/23337486.2015.1006879.

Bateson, G. (1972). *Steps to an ecology of mind* (pp. 177–193). Chicago: University of Chicago Press.

Bawden, R. (1992). Evaluation – Learning through dialogue. In Agpak (Ed.), *LS1: Essays in sabbatical reflection*. Hawkesbury, Richmond: University of Western Sydney.

Bawden, R. J. (1997). The community challenge: The learning response. Invited plenary paper at the Annual conference of the community development society, Athens, Georgia, USA, August 1997. p 8. Online at: https://www.decd.sa.gov.au/sites/g/files/net691/f/the_community_challenge_the_learning_response.pdf.

Bawden, R. J. (2010). Messy issues, worldviews and systemic competencies, chapter 6. In C. P. Blackmore (Ed.), *Social learning systems and communities of practice* (pp. 89–102). London: Springer.

Bawden, R. J., & Packham, R. G. (1993). Systemic praxis in the education of the agricultural practitioner. *Systems Practice, 6*, 7–19.

Bergin, A., & Townsend, J. (2007). *A change in climate for the Australian Defence Force*. Special report issue 7, Australian Strategic Policy Institute, Canberra, Australia. Online at: https://s3-ap-southeast-2.amazonaws.com/ad-aspi/import/SR7_climagechange_ADF.pdf?YwRw5so1hx9XuPgkYzzeNLtkaEJ9ooWQ [downloaded 13 August 2011].

Boisot, M., & McKelvey, B. (2011). Complexity and organization-environment relations: Revisiting Ashby's law of requisite variety. In P. Allen, S. Maguire, & B. McKelvey (Eds.), *The SAGE handbook of complexity and management* (pp. 279–298). London: Sage.

Bonaparte, N. (1804). Letter to the King of Prussia. In C. J. Herold (Ed.)., (2016) *Mind of Napoleon: A selection of his written and spoken words*. Sevenoaks: Pickle Partners Publishing.

Boxall, S. (2012). *Complexities, uncertainties and climate securitisation-Tools for exploration*. Headquarters New Zealand Defence Force presentation to the Global change and energy sustainability forum 2012, Australian Department of Defence, Canberra.

Burns, D. (2007). *Systemic action research: A strategy for whole system change*. Bristol: Policy Press.

Churchman, C. W. (1971). *The design of inquiring systems*. New York: Basic Books.

Cleugh, H., Stafford Smith, M., Battaglia, M., et al. (Eds.). (2011). *Climate change: Science and solutions for Australia*. Collingwood: CSIRO Publishing.

CNA. (2007). *National security and the threat of climate change*. The Military advisory board, The CAN Corporation, Alexandria Virginia, USA. Online at: https://www.cna.org/CNA_files/pdf/National%20Security%20and%20the%20Threat%20of%20Climate%20Change.pdf [downloaded 13 August 2011].

Collins, J. M. (1998). *Military geography for professionals and the public* (First ed.). Dulles: Brassey.

Crutzen, P. J. (2006). The 'Anthropocene'. In E. Ehlers & T. Krafft (Eds.), *Earth system science in the anthropocene: Emerging issues and problems* (pp. 13–18). Berlin: Springer.

Department of Prime Minister and Cabinet. (2016). *Australia's cyber security strategy*. Commonwealth of Australia. Online at: https://cybersecuritystrategy.pmc.gov.au/ [Accessed 3 July 2017].

Dupont, A., & Pearman, G. I. (2006). *Heating up the Planet: climate change and security*. Lowy Institute paper 12, Lowy Institute, Sydney, Australia. 143pp.

Durrant, C., & Holloway, J. (2013). Global change impacts on defence. Presentation to the Pacific Environmental Security Forum, Sydney, Australia, 16–19 April. Summary available online at: http://www.pesforum.org/docs/2013/3_AUS_PESF_presentatioen_summary.pdf [Accessed 17 Nov 2017] p 9, Day 4.

Ehlers, E., & Krafft, T. (Eds.). (2006). *Earth system science in the Anthropocene: Emerging issues and problems*. Berlin: Springer.

Einstein, A. (1946). Atomic education urged by Einstein. *New York Times*–25 May 1946.

Galgano, F. A. (2017). Military Geography. Oxford Bibliographies. Doi https://doi.org/10.1093/OBO/9780199874002-0029 Online at: http://www.oxfordbibliographies.com/view/document/obo-9780199874002/obo-9780199874002-0029.xml [last modified 27 September 2017, Accessed 19 Nov 2017].

Galgano, F. A., & Palka, E. J. (Eds.). (2011). *Modern military geography*. New York and London: Routledge.

Grant, T. J. (2014). On the military geography of cyberspace. In Liles S (ed) proceedings, 9th International conference on cyber warfare & security (ICCWS 2014), Purdue University, West Lafayette, Indiana, USA, 24–25 March 2014. pp 66–76.

Harmon, R. S., Dillon III, F. H., & Garver Jr, J. B. (2004). Perspectives on military geography–the military operating environment. In D. R. Caldwell, J. Ehlen, & R. S. Harmon (Eds.), *Studies in military geography and geology* (pp. 7–20). Dordrecht: Kluwer Academic Publishers.

Harvey, F. (2006). Ontology. In B. Warf (Ed.), *Encyclopedia of human geography* (pp. 341–343). Thousand Oaks: SAGE.

Holloway, J. (2012). Impacts on defence preparedness of changes in strategic military geography: Outcomes of the 8 October 2011 Initial scoping workshop. Client report DSTO-CR0–2012-0346, Defence Science & Technology Organisation, Canberra, Australia.

Holloway, J., Thomas, M. D., & Durrant, C. (2014). Strategic military geography: Climate change adaptation and the military. In W. L. Filho (Ed.), *Handbook of climate change adaptation*. Berlin: Springer.

Homer-Dixon, T., Walker, B., Biggs, R., et al. (2015). Synchronous failure: The emerging causal architecture of global crisis. *Ecology and Society, 20*(3), 6. https://doi.org/10.5751/ES-07681-200306.

IPCC. (2013). Fifth assessment report: Climate change 2013-Working group 1 report "The Physical Science Basis". Inter-governmental panel on climate change, Geneva, Switzerland. Online at ipcc.ch/publications_and_data/publications_and_data_reports.shtml [Accessed 20 Sept 2013].

Ison, R. (2010). *Systems practice: How to act in a climate change world.* London: Springer.

Ison, R., Grant, A., & Bawden, R. (2010). *Scenario praxis for systemic and adaptive governance: A critical review.* VCCCAR scenarios for climate adaptation working paper, Victorian Centre for Climate Change Adaptation Research, Melbourne, Australia.

Kilcullen, D. (2016). *Blood year: Islamic state and the failures of the war on terror.* Carlton, Australia: Black Inc.

Kitchener, K. (1983). Cognition, meta-cognition and epistemic cognition: A three level model of cognitive processing. *Human Development, 26*, 222–232.

Kitchin, R. M. (1998). Towards geographies of cyberspace. *Progress in Human Geography, 22*(3), 385. https://doi.org/10.1191/030913298668331585 1998 22: 385.

Kvanvig, J. (2003). *The value of knowledge and the pursuit of understanding* (p. 5). Cambridge: Cambridge University Press.

Laksmana, E. A. (2011). Double jeopardy: Climate insecurities and their implications for asian armed forces. *Defence Studies, 11*(2), 271–296.

Megginson, L. C. (1963). Lessons from Europe for American business. *Southwestern Social Science Quarterly, 44*(1), 4.

Payne, M., Senator the Honourable. (2017). *Capability to fight cyber criminals.* Press release by the Prime Minister, Minister for Defence and Minister Assisting the Prime Minister for Cyber Security, 30 June 2017. Online at: https://www.minister.defence.gov.au/minister/marise-payne/media-releases/prime-minister-minister-defence-and-minister-assisting-prime [downloaded 3 July 2017].

Press, A., Bergin, A., & Garnsey, E. (2013). *Heavy weather: Climate and the Australian Defence Force.* Special report issue 49, Australian Strategic Policy Institute, Canberra, Australia. Online at: https://www.aspi.org.au/publications/special-report-issue-49-heavy-weather-climate-and-the-australian-defence-force [downloaded 15 April 2014].

Raworth, K. (2012). A safe and just space for humanity: Can we live within the doughnut? Oxfam discussion paper. Online at: https://www.oxfam.org/sites/www.oxfam.org/files/dp-a-safe-and-just-space-for-humanity-130212-en.pdf [Accessed 19 Nov 2017].

Reid, W. V., Chen, D., Goldfarb, H., et al. (2010). Earth system science for global sustainability: Grand challenges. *Environment and Development, 330*, 916–917.

Rescher, N. (2005). *Value matters: Studies in Axiology.* Frankfurt: Ontos Verlag.

Rockström, et al. (2009). Planetary boundaries: Exploring the safe operating space for humanity. *Ecology and Society, 14*(2), 32. [online] URL: http://www.ecologyandsociety.org/vol14/iss2/art32/ [Accessed 18 Nov 2017].

Röling, N. (2002). Emerging issues and challenges for FFS: An introductory overview. Paper for the international learning workshop on Farmer Field Schools (FFS): Emerging issues and challenges, Yogyakarta, Indonesia, 21–25 October.

Senge, PM (1990) *The fifth discipline: The art & practice of the learning organization.* Doubleday Business, New York, USA

Senge, P. (1994). *The fifth discipline fieldbook: Strategies and tools for building a learning organization.* New York: Bantam Doubleday Bell.

Singer, P. W., & Friedman, A. (2014). *Cybersecurity and cyberwar: What everyone needs to know.* New York: Oxford University Press.

Smith, J. E. (translator) (1956). Kant's Weltanschauung by Richard Kroner. Authorised English translation of Kant's Weltanschauung (Tubingen, 1914), University of Chicago Press, Chicago, Illinois, USA.

Snowden, D. J., & Boone, M. E. (2007). A leader's framework for decision making. *Harvard Business Review* November.

Steffen, W., Crutzen, P. J., & McNeill, J. R. (2007). The Anthropocene: Are humans now overwhelming the great forces of nature? *Ambio, 36*(8), 614–621.

Taylor, G. (1963). Geographers and world peace–a plea for geopacifics. *Geographical Research, 1*(1), 3–17. https://doi.org/10.1111/j.1467-8470.1963.tb00002.x.
Templeman, D., & Bergin, A. (2008). *Taking a punch: Building a more resilient Australia.* ASPI Strategic Insight, Australian Strategic Policy Institute, Canberra, Australia. Online at: https://s3-ap-southeast-2.amazonaws.com/ad-aspi/import/SI39_Taking_a_punch.pdf?vcIJUhU3L9HG1gmbj04jrpjQFW3YIOUj [Accessed 27 Nov 2017].
Thomas, M. (2012). Assessing the risk of global climate change on the ADF. *e-International Relations* 8 March 2012. Online at: http://www.e-ir.info/2012/03/08/assessing-the-risk-of-global-climate-change-on-the-australian-defence-force/ [downloaded on 10 September 12].
UNSCR. (2000). Resolution 1325. Adopted by the Security Council at its 4213th meeting, 31 October. Online at: http://www.un.org/ga/search/view_doc.asp?symbol=S/RES/1325(2000) [Accessed 19 Nov 2017].
US Navy. (2010). Navy climate change roadmap. Document Reference SER NO9/10U103021 Online at: http://www.navy.mil/navydata/documents/CCR.pdf [downloaded 4 July 2011].
Warf, B. (2001). Segueways into cyberspace: Multiple geographies of the digital divide. *Environment and Planning B: Planning and Design, 28*, 3–19. https://doi.org/10.1068/b2691.
Woodward, R. (2004). *Military geographies*. Oxford: Wiley.

Chapter 5
Climate Change as a Hyperthreat

Elizabeth G. Boulton

1 Introduction

> Our increasing knowledge of global warming ends all kinds of ideas, but it creates other ones (Morton 2013 p.127).

The collapse of the Copenhagen climate talks in 2009 may come to be regarded as a historical marker for the time when it became clear that evidenced-based science, on its own, could not secure a pathway to a safe climate. It was now evident that there were substantial socio-economic, political, cultural, and psychological barriers at play (Bodansky 2010; Dubash 2009). Although a grim realisation, confronting this truth led to an intellectual pivot in climate policy research. There was an intensification and expansion of research into effective science communication and related behaviour change, and this effort allowed some common themes and conclusions to emerge.

Insights from disciplines as diverse as neuroscience, linguistics, and philosophy converge in identifying the criticality of *deep frames*; world views that are hardwired into human neurocircuitry over a lifetime and influence decision-making and behaviour, mostly at the subconscious level (Nisbet 2009; Lakoff 2010; Di Leo 2013; Palsson et al. 2013). *Moderns*—that is, people born in the industrial era but before the time of significant global warming impacts—have a particular way of thinking (a deep frame) that may hinder their ability to respond effectively to the threat of environmental and climate change. Moderns struggle to conceive of dependency upon the natural world. An overview of climate framing research finds that while it is now possible to identify four dimensions to a successful frame, (affective,

E. G. Boulton (✉)
Fenner School for Environment and Society, Australian National University, Linnaeus Way, Canberra, ACT 2601, Australia
e-mail: Elizabeth.boulton@anu.edu.au

S. Pearson et al. (eds.), *Australian Contributions to Strategic and Military Geography*, Advances in Military Geosciences,
https://doi.org/10.1007/978-3-319-73408-8_5

narrative, conceptual, and philosophical)—a solution remains elusive (Boulton 2016a).

Accordingly, the next clarion call has been for a new way to frame climate change—a new conception, philosophy, or world view—that might liberate people from dated constructs. A bold contribution has been made by eco-philosopher Timothy Morton, who reframes global warming as a *hyperobject*; something 'massively distributed in time and space relative to humans' (Morton 2013 p.1). Morton's work has attracted both acclaim and controversy; nonetheless, its creative vitality refreshes the discussion and invites further development.

In this chapter, Morton's hyperobject conception is considered in the context of climate-security discourse. Specifically, his hyperobject theory is used to develop the alternate notion of a *hyperthreat*. The approach taken differs from prior climate-security discourse in two ways. Firstly, it is allegorical, creative, and philosophical—which is in keeping with the insights already developed in climate-framing research (Boulton 2016a). Secondly, it departs from the long-established delineation in climate-security literature, whereby considering how to respond to climate change is regarded as solely a civil responsibility, where military commentary is unwelcome. Here, in contrast, conceptual approaches to *threat* which are found within traditional security studies (those focused on present-day national security and defence policy, for example), are applied to the threat of climate change. The two selected were developed by the war-theorist Carl von Clausewitz (Clausewitz 1832). Firstly, the *character* of the hyperthreat is considered—this is a broad type of threat analysis, which could also be considered a framing activity on its own. Secondly, von Clausewitz's notion of *moral forces* is explored, primarily through focus upon just war theory (JWT).

To progress this enquiry, the chapter begins with the theoretical and conceptual background: an overview of prior climate-security literature and of Morton's hyperobject frame, with some critique. The hyperthreat concept is then introduced, and its dimensions and implications are explored. Overall, the approach is to twist and play with Morton's metaphor a little, as though turning a kaleidoscope, to see what further insights may emerge. While this involves disputing and countering some of Morton's conclusions, it is done with awareness that if Morton had not created the metaphor in the manner he did, these new discussions would not have been possible.

2 Theoretical Background

2.1 Environmental Security Discourse until Now

Hyperthreat exploration occurs in the context of the vast field of environmental security studies, which began in 1971 (Barnett 2001) and now incorporates numerous well-established specialties such as food, water, soil, energy, and climate security. This research activity has been paralleled by related theoretical reconceptualisation. Notably, the concept of *human security*—freedom from want, hazard, or fear—(Brauch 2002) has been ascendant. Human security has been progressively reworked into related thinking like the quartet of peace, security, development and environment approach (Brauch 2008) or the human, gender, and environmental security (HUGE) notion (Spring 2009). Other explorations include the idea of greening the UN Security Council (Penny 2007), existential survival frames such as *planetary boundaries* (Rockström et al. 2009), and the expanding notion of *environmental peacekeeping*, whereby the critical need for solutions propels constructive alliances and peaceful arrangements (Rüttinger et al. 2015; Tänzler et al. 2010).

Focusing upon climate security, this discourse is best encapsulated in the Intergovernmental Panel on Climate Change's (IPCC) most recent assessment (IPCC 2014). Although criticised for inconsistency (Gleditsch and Nordås 2014), and despite ongoing debates about methods and subsequent conclusions (Salehyan 2014; Buhaug et al. 2014; Hsiang and Burke 2014), the IPCC's general conclusions on climate security are accepted, and inform analysis in this chapter. Briefly, that is, that 'human security will be progressively threatened as the climate changes… [which] emerges from the interaction of multiple factors' (Adger et al. 2014 p.758). The IPCC also regard the potential for violent conflict as an 'emergent risk because factors such as poverty and economic shocks that are associated with a higher risk of violent conflict are themselves sensitive to climate change' (Oppenheimer et al. 2014 p.1042).

Within 'traditional security' discourse, climate is routinely considered alongside a wide range of other factors including organised crime, health epidemics, cyber warfare, space, and even genome editing (Clapper 2016). Echoing the broader trend in climate policy of calling for a new grand narrative or guiding frame, traditional security analysts have also argued that the arrival of climate change demands that the notion of security be utterly reconceived (Mathews 1989; Dupont and Pearman 2006). Aside from periodic warnings made by various high-profile defence personnel (Hannam 2015; CNA MAB 2014), the approach of traditional security agencies is best illustrated by *what they do*—their formal positions clarified in nation-level security and defence strategies. Generally, this involves militaries planning to conduct increased humanitarian assistance and disaster relief (HADR) functions and, like most other sectors of society, undertake internal climate mitigation and adaptation activities (Thomas 2013). The term *hyperwar* has been used to describe future warfare involving artificial intelligence and machine cognition, in which humans are absent from physical conflict spaces; however, it is not clear whether this draws upon Morton's hyperobject conception (Allen and Husain 2017).

The *climate securitisation* frame and the idea of involving the traditional security sector in the climate policy field creates unease among some analysts, as it presses against long-observed boundaries of what constitutes a civil society governance issue and what represents a military issue. Additionally, a climate securitisation frame does introduce risk. It can result in simplistic and sensationalised interpretations. For example, Hulme argued that when violent conflict in Darfur, Sudan was attributed to climate change, it served to 'divert attention from deeper, political factors' (Hulme 2007 p.26). Barnett notes that panoptic visions of the world serve the traditional security sector's interests, and worries that such framing may also be used to 'justify crude interventions to control local places and peoples' (Barnett 2009). Many note that *tools of force* or traditional security methods are utterly mismatched to dealing with climate change (Dabelko 2009). Another risk is that a climate security frame may chill cooperation and sideline planning for peaceful solutions (Floyd 2008). The entire idea of identifying climate change as a *threat multiplier* has been challenged through a critique of the causes of the Syrian Civil War (Selby et al. 2017). Selby and others argued that widespread commentary in security, political, and media discourse linking the Syrian war to climate change was based upon flawed analysis, which ignored the significant economic and political dimensions. This case study is an example of the need for greater caution around *climate securitisation*.

Outside climate-security academic literature, it should be acknowledged that framing global warming as a threat is not unique; many have conceived that environmental threats require warlike mobilisation. For example, Woodbridge defined ecological decline as the new enemy, and articulated a top-down globalised response in which 'the war must be waged on the battlefields of technological innovation' (Woodbridge 2004 p.265). Lester Brown's various Plan Bs offered large-scale, Marshall Plan-inspired responses (Brown 2008). Several other grand or meta-scale narratives explore how mass transformation is needed or could occur (Lovelock 2009; Gilding 2011; Diamond 2005; Klein 2014). The prevalence of apocalyptic climate narratives in film and literature (Swyngedouw 2010; Nielsen 2017)—perhaps prompted by the Pentagon's early inquiries into this area (Schwartz 2003)—also highlights awareness of the warlike destruction threatened by global warming. However, while environmental threats have previously been conceived in warlike mass-mobilisation terms, the fact that the necessary large-scale response has not yet occurred points to the effectiveness of deep framing barriers (among other issues) discussed earlier and analysed in more detail by Boulton (Boulton 2016a).

Given the existing critique and hesitancy about *climate securitisation* discourse, the decision, in this chapter, to draw upon traditional military concepts and analytical tools to consider climate change requires some explanation. Warfare has been a part of the human experience since antiquity; it is an inherently violent and chaotic activity. As von Clausewitz described, it is 'the province of danger…suffering and chance [where] the horror of its elements excites repugnance' (Clausewitz 1832 pp.102–140). Accordingly, extensive military literature exists on how best to prepare for—and respond to—hostile threats. Although the differences between warfare and environmental threats are profound, nonetheless, it may be that through grappling with the idea of *threat* in theory and practice over millennia, the military profession has intellectual gifts to offer.

The idea of an intellectual *gift consciousness* draws upon already established notions of the gift economy (Cheal 2015) and gift creativity (Hyde 2007). It asks, what *gifts* (skills, knowledge, expertise) belong to different sectors of society, and how can they be best applied to the climate change problem? Intellectual gift consciousness may help progress transdisciplinary research, overcome siloed thinking, and even help defuse intellectual turf wars. In this chapter, crossing the boundary of what military thought is expected to grapple with, while fusing insights from various other disciplines, is done with the intention of equipping the analyst with the best cognitive tools for the task at hand. It does not imply application of traditional military-style solutions; rather, in the spirit of a pioneer, it seeks merely to consider what emerges when a problem is viewed through another lens. This is undertaken as, according to the preceding body of research on climate framing, it is present-day thought constructs themselves which are the key hurdle that needs to be overcome, to enable an effective response.

2.2 *The Hyperobject Frame: An Overview*

The historic moment at which hyperobjects become visible by humans has arrived. This visibility changes everything...This is a momentous era (Morton 2013 p.128).

In his book, *Hyperobjects, Philosophy and Ecology after the End of the World* (Morton 2013), Morton describes a hyperobject as a type of *thing* that operates across such vast time and spatial dimensions that it cripples the ability of humans to either conceive of or respond to it. However, the hyperobject is not a literal thing; rather, it needs to be understood as a grand metaphor used by Morton to convey a new philosophical outlook (Fig. 5.1). The theoretical building blocks which underpin Morton's work include Heideggerian philosophy, object-oriented ontology (OOO), and *new materialism*; all of which de-centre the human being as the locus of meaning and importance in the universe. Hyperobject framing sees a new world conceptual order in which the human is only one object among trillions of living and inanimate objects, such as rocks, chairs, or micro-bacteria, all of which have equal existential status. This is in contrast to the Anthropocene narrative, in which human prominence is emphasised by the very naming of the era, (*anthropo* means *human*), and where the discursive focus is upon human power over the natural world (Crutzen 2006; Steffen et al. 2007).

Pertinent to the climate-security discussion is Morton's notion that the hyperobject terrorises people internally. Humans can no longer shape their worlds; rather, their future existence involves being stuck in something they cannot even understand. They will be buffeted about by climate change, in the future, like leaves in the wind. As humans tinker with old economic and governance systems to try to manage the problem, they delude themselves that they can control something that is already in train and beyond their capacity to influence.

In confronting the challenge of explaining something entirely new—addressing the *unavailability heuristic* that may hinder climate communication (Lazarus 2009)—Morton's answer is to use layers of metaphors as his cognitive building blocks. Thus, his approach is allegorical and imaginative. An overview of the hyperobject conception, with key discursive terms and examples of Morton's metaphor use, is shown in Fig. 5.2.

Fig. 5.1 Cocoon Shield by Danish sculptor Jens Galschiøt (Galschiøt 1991). Eco-philosopher Timothy Morton argued that humans must break through the limitations of modern-era thought constructs, if they are to properly perceive their new hyperobject reality

In this chapter, Morton's hyperobject conception is developed further to consider that climate and environmental change represents a new sort of threat—a hyperthreat. Countering or minimising the destructive power of the hyperthreat is dependent upon the ability of humans to effect a philosophical pivot; to cognitively upgrade to their new era.

The effectiveness of the hyperobject metaphor as a way of framing climate change has been analysed in depth elsewhere (Boulton 2016a). This finds that Morton's narrative style does accord with many of the insights from climate framing research, and that the hyperobject frame is powerful in helping humans to confront the full scale of their predicament. Humans' existing structures, institutions, and problem-solving methods are revealed as not only trivial in scale to the hyperobject, but also mismatched towards the true nature of the problem. This provides a key insight—the hyperobject demands a hyper-response; a solution that is not moored to existing institutional, governance, or conceptual practices.

However, there are two areas in which the hyperobject frame breaks down. The first is the association of vulnerability with shame. This idea echoes hegemonic masculinity constructs (Connell and Messerschmidt 2005; Courtenay 2000) and is disputed by vulnerability researchers who argue vulnerability is not a shameful condition, but rather an essential part of being human, with complex links to positive emotions such as courage, passion, authenticity, and the capacity to build trust (Nussbaum 2003; Brown 2012). Secondly, several commentators argue that

HYPEROBJECT: FIVE CHARACTERISTICS[a]

1. ***Viscosity*** – Like fog, it is everywhere. Like honey—it sticks to humans. It changes form as humans interact with it. '*Haunt(s) my social and psychic space*' (p. 36).
2. ***Non-locality*** – Distributed across vast geographical areas, but also operates at the quantum particle level. Cannot be seen at human, local levels. Disables humans' normal cause-and-effect calculations.
3. ***Temporal undulation*** – Operates across planetary—not human—time frames. May defy human gravity-time-space conceptions. '*Impossible to hold in mind*' (p. 58).
4. ***Phasing*** – Like Lorenz's butterfly diagram, which helps to describe chaotic systems, hyperobjects are multi-phased. Humans cannot tell which phase state they may be seeing.
5. ***Interobjectivity*** – Operates through other objects and is never seen outright. '*We never see the hyperobject directly. We infer it from graphs, instruments, tracks in a diffusion cloud chamber, sunburn... we see the shadows of [it]*' (p. 153).

HYPEROBJECT'S ARRIVAL: THE IMPLICATIONS

New ontological viewpoint. Humans are forced to understand their existence in a fundamentally new way:

- Humans are demoted from their role as (perceived) rulers of Earth to merely one type of object, among many, that must all coexist.
- Humans lose agency—the hyperobject is now the main shaping agent on Earth.
- Humans' new status is weak, lame, and vulnerable.

Hyperobjects terrorise. The hyperobject affects humans, not only through physical changes on Earth, but also at a deep existential and psychological level:

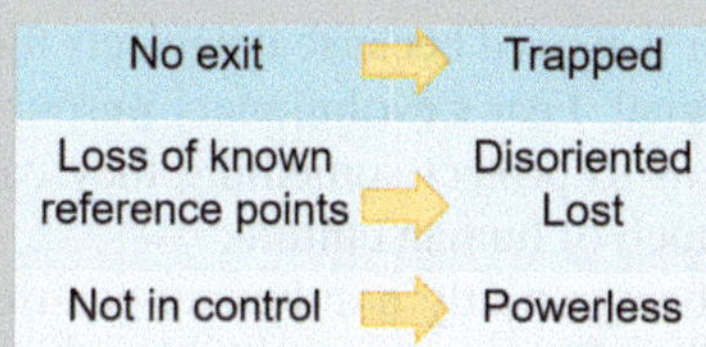

We have woken up inside an object, like a movie about being buried alive...a time of hypocrisy where every decision is 'wrong' (p. 160).

Two hundred years of seeing humans at the centre of existence, and now the objects take revenge, terrifyingly huge, ancient, long-lived, threateningly minute, invading every cell in our body (p. 155).

Fig. 5.2 Hyperobject: Characteristics and implications

[a]All page references to: Morton T (2013) *Hyperobjects, Philosophy and Ecology after the End of the World*. Minneapolis, University of Minnesota Press

the hyperobject frame is too pessimistic and grim—that its harsh outlook fails to inspire people to take positive or creative actions (Leonardi 2014; Daggett 2014; Holmes 2012).

3 How and why the Hyperthreat Departs from the Hyperobject Frame

The hyperthreat conception adopts the five characteristics of Morton's hyperobject (see Fig. 5.2) but, perhaps stubbornly or regressively, departs from Morton's theory by maintaining belief in some human agency and ability to influence global warming. It accepts the advice of the IPCC that humanity still has the capacity to influence the magnitude and speed at which global warming occurs (IPCC 2014; Hansen et al. 2013). However, there is an urgency in this task; the longer humans delay energy transformation, the exponentially harder it becomes. Thus, speed is a critical part of a necessary response, yet the hyperobject notion of the powerless human does not allow this aspect to be brought to the fore. Related to the issue of agency is the question of how to respond to the hyperobject. Morton encourages humans to coexist with the hyperobject. The predicted severe impacts of dangerous global warming are treated as something to endure—of no more consequence than an ant being trampled underfoot by a pedestrian on a sidewalk.

The hyperthreat notion differs from Morton's view, in that while it also acknowledges human vulnerability, it does not equate this with shame, but instead argues that awareness of human's new vulnerability provides an ethical and survival-based motive to act. Evolutionary psychologists posit that fear is a pragmatic and useful emotion that alerts the human body to the need to take action (Tooby and Cosmides 2008). Accordingly, others have argued that the problem may actually be a fear-deficit; people do not understand the danger they face (Loewenstein and Schwartz 2010; Loewenstein 2010). So, while Morton's approach of helping readers to acknowledge their vulnerability and connect with associated feelings of fear has value, his approach to fear is flawed because it neglects the next step—how humans can respond to the fear signal. Fear's evolutionary purpose is to alert and prime the human to take action so as to protect something they care about; this is the *hero story* that permeates so much of human culture.

Morton's concept, in being strictly non-human-centric, risks dissolving or rendering inconsequential two of the most powerful human attributes: hope and courage. The word courage derived from the Latin word *cor*, for heart—is an affective state, and this points to the importance of the emotional dimensions of engagement with the hyperthreat. While attributes such as courage are not prominent in climate response literature, traditional military strategists view the affective dimension as a crucial part of confronting threats. Von Clausewitz regarded *moral forces*—which he described as being the spirit and feeling of a force, its capacity for courage and boldness— as being 'amongst the most important subjects in War':

> *These forces fasten themselves soonest and with the greatest affinity on to the Will which puts in motion and guides the whole mass of powers, uniting with it as it were in one stream, because this is a moral force itself. Unfortunately they will escape from all book-analysis, for they will neither be brought into numbers nor into classes, and require to be both seen and felt* (Clausewitz 1832 p.251).

Hence, in contrast to Morton's vision, the hyperthreat conception assumes that it is possible for humans to tame the hyperthreat, but notes that doing this may rely heavily upon moral forces.

Finally, the hyperthreat notion incorporates other aspects of environmental degradation, which inevitably interweave with and impact upon the way global warming occurs, such as is encapsulated in the Rio Declaration on sustainable development (UNEP 1992); the *planetary boundaries* conception (Rockstrom et al. 2009), and broader IPCC work on climate impacts that considers other factors that bear upon environmental resilience. However, for brevity, these aspects of the hyperthreat are not further elaborated upon in this chapter.

3.1 Why Threat?

For general climate security policy participants, who may be unfamiliar with the nuances of *new materialism* and 'vibrant matter' (Bennett 2009), the word *object* may seem bland and forgiving. Thus here, for clarity, *object* has been replaced with *threat* to more accurately portray the harm and suffering that global warming is expected to deliver, as articulated in IPCC climate impact reports (IPCC 2014), and special reports on extreme weather (IPCC 2012). The word *threat* also reflects the intention to utilise traditional military approaches. This activity of threat analysis—correctly identifying the nature of a threat—is a fundamental component in military strategy. As von Clausewitz explained, wars differ in character, and understanding the unique nature of each threat is 'the first, the grandest, and most decisive act of judgement which the Statesman and General exercises…[it is] the most comprehensive, of all strategical questions' (Clausewitz 1832 p.121). This task, of correctly defining the character of a war or threat remains difficult and paramount, with present-era military generals lamenting that this was the principal failure of Western interventions in the Iraq/Afghanistan wars (Day 2015). As this step is so important, it is typically supported by sophisticated intelligence collection and multidisciplinary analysis processes (US DOD 2011). Acknowledging the cruciality of this step, attention will now turn to consider the character of the hyperthreat.

4 Hypothesis: The Hyperthreat Frame

What type of threat is the hyperthreat? What are its dimensions? To start this recalibration, one relevant idea is Galtung's structural violence concept (Galtung 1969), in which uneven political power and resource access causes harm. This has previously been applied in an environmental context by Barnett (Barnett 2008; Barnett 2007). Related to this, is Rachel Carson's holistic view of how pesticides harm wildlife, and in turn, human beings—if a wider multi-disciplinary and temporal analytic mode is used (Carson 1962; DeMarco 2017).

4.1 Slow Violence

Associated with both Galtung and Carson's macro-analytical approaches, but in keeping with deep framing and metaphoric constructs, is Rob Nixon's concept of *slow violence*, defined as:

> Violence that occurs gradually and out of sight, a violence of delayed destruction that is dispersed across time and space, an attritional violence that is typically not viewed as violence at all (Nixon 2011 p.2).

Nixon's work does already inform that of Morton; however, it is valuable enough to be drawn out further. Nixon's main argument relates to economic and industrial activities that degrade local environments and thereby harm population groups in layered and gradual ways. Examples of slow violence he discusses include the Bhopal gas disaster in India, oil extraction in the Ogoni area of Nigeria, the construction of mega dams in India, and the use of landmines and cluster bombs in Angola and Afghanistan. Such activities, he argues, can remove livelihoods, damage agricultural production, erode water quality, and displace people; thus, they slowly harm and kill. Nixon quotes Nobel Peace Prize winner Wangari Maathai of Nairobi to show that this conception of violence—while it may be alien to wealthy, modern countries—accords with those who more directly rely upon their natural environments:

> Losing topsoil should be considered analogous to losing territory to an invading enemy. And indeed, if any country were so threatened, it would mobilize all available resources, including a heavily armed military, to protect the priceless land (Wangari Maathai, 2008; cited in Nixon 2011 p.130).

Nixon also briefly applies the idea to climate change, through which people stand to suffer similar but larger-scale forms of slow violence. It is proposed here that slow violence is part of the hyperthreat's modus operandi.

4.2 Irreversibility

Already long-articulated in climate and environmental policy (UNEP 1992), the issue of irreversible destruction must be part of the hyperthreat conception. In the case of human wars—even with nuclear, biological and chemical warfare—the possibility of rebuilding and effecting some remediation exists. However, for climate change, the calculus is far more serious; restoration may not be possible. There is a finality and savagery in runaway, dangerous climate change impacts—such as widespread extinctions or acidification of the ocean—that distinguishes them from traditional military threats.

4.3 Enemy, Culpability, and Saboteurs

Morton's hyperobject frame and its non-human-centric view does not linger on the issue of culpability; an approach that has advantages in what is already a heavily polarised debate (Kahan et al. 2011; Bliuc et al. 2015). While it is obvious that many people inevitably contribute to global warming through participation in society, this *indirect culpability*—with its innumerable *climate equity* delineations (Heyward 2007; Klinsky et al. 2017)—needs to be distinguished from *direct culpability*.

There is now a substantial body of research that demonstrates how special interest groups deliberately seek to derail climate response—they fight against regulatory measures and public information approaches to limit greenhouse gas emissions (Oreskes and Conway 2010; Lewandowsky et al. 2015; Klein 2014; Roy 2014; Hamilton 2007; Hoggan and Littlemore 2009; Dunlap and McCright 2011). These undermining activities are regarded by many as legitimate participation in economic-socio-political democratic processes. Hence, although deliberate sabotage occurs; thus far, the primary means by which civil society has addressed this is through informal actions. Generally, this involves community members or organisations, (like 350.org) drawing public attention to an accused company, thereby risking damage to the company's reputation and possibly influencing its social licence to operate. No formal measures are in place to address this problem, bar some tentative discussions, such as within the United Nations (UNFCCC 2017; Tabuchi 2017; Kemp 2017).

Aside from conscious attempts to slow institutional and governmental responses to address global warming, within the category of direct culpability there may be a further, more opaque category of culpability, which has elsewhere been described as *strategic ignorance*—situations in which threat or risk are ignored or minimised due to a combination of factors such as focus upon profit, self-interest, stereotypes, social taboos, or academic disciplinary constraints (McGoey 2012). Strategic ignorance could be viewed as an alternative way of conceiving and considering the framing question.

4.4 A Hazard or a Threat?

When the issue of direct culpability and global warming is considered within traditional security and climate-security literature, the enquiry is stymied by extant word definitions. Literature in both these areas generally adopts definitional norms within geography and disaster risk literature, where hazards are differentiated from military threats as not having a brain or a conscious hostile intent (Blaikie et al. 2014). While this logic did suit a pre-climate era—naturally occurring cyclones or tsunami were not caused by an entity with a conscious intent —the distinction is no longer as neat. The improving ability of climate attribution science to statistically link extreme events to a warmer climate (Parmesan and Yohe 2003; Stott et al. 2016) is part of the improving evidence base, and this needs to inform the way in which such hazards or threats are perceived.

Traditional thought and legal constructs would not link climate saboteurs with dangerous climate impacts, yet this old cause-and-effect arrangement does not capture the new reality or Morton's *non-locality* characteristic of hyperobjects. This presents an example in which existing human mechanisms are, as Morton described them, as 'useless as the proverbial chocolate teapot' (Morton 2013 p.103). In Beck's terms, they are *zombie* institutions and thought constructs—ones which are 'dead even though they appear to still be functioning' (Boyne 2001 p.47). They are considered *dead* because they no longer interact in a *lively* or responsive way to the changing external world. Old logic on culpability, enemy, and hostile intent to harm must be reconceived in such a way that allows humanity new mechanisms to protect itself in a climate-era.

4.5 Victims Are Mostly Silent or out of View

The hyperthreat's main victims are currently voiceless—the young; future generations; and the quietly crumbling and disintegrating great ecosystems of the world with their respective unique civilisations of mammals, birds, plants, and invertebrates. The hyperthreat enemy rarely sees or hears from its victims; it cannot hear their cries which might evoke empathy. Like Arendt's analysis of *the banality of evil*, decision makers are so removed from the cause-and-effect impact of their actions that they risk abdicating their *humanness*, which allows ethical judgements to occur (Arendt 1963). The hyperthreat performs a type of remote killing and destruction that makes the drone warfare ethics discussion seem simple in comparison. The *separation factor* between drone operator and victim has been the subject of much deliberation (Brunstetter and Braun 2011; Chamayou 2015)—will it inhibit some of the moral safeguards that guide a person to only kill when essential and as a last-resort mechanism? This type of thinking needs to be applied to the hyperthreat, which has a far greater separation factor.

5 Some Implications of the Hyperthreat Framing

5.1 Hyperthreat and Reasons to Fight

If we accept that the hyperthreat has warlike destructive capabilities (or worse), and the necessary mass mobilisation and response eludes humanity at this point, it could be helpful to investigate why societies have chosen to go to war in the past.

In an analysis of warfare from antiquity to the present, Chaliand found an overwhelming number of reasons why humans go to war, including procuring slaves, obtaining resources, ideology, establishing religious dominance, and the desire to amass power, among others (Chaliand 1994). Given the huge range of reasons and contexts, it is pragmatic to narrow this inquiry to *just war theory* (JWT). JWT is a catchphrase for the study of war ethics from ancient times (e.g., by Confucius, Aristotle, Plato, and Mencius), to theories that introduced the idea of nations after the Peace of Westphalia in 1648, to more recent deliberations on the ethics of pre-emptive actions to address terrorism (Fotion 2007). This body of knowledge has been distilled into common principles used to inform modern-day warfare deliberations. As Fotion explained, the choice to go to war requires a departure from a range of absolute principles held in common by many societies (do not kill, etc.); thus, there must be exceptionally *good reasons* to do so. Three such *good reasons* used in the past apply to the hyperthreat: general destruction, loss of autonomy, and basic survival considerations.

Within JWT, the risk of general destruction is regarded as the first overriding good reason why people would collectively rally against an aggressive force:

> Once a nation has been taken over by another, the prostrate country can be exploited, plundered, and even totally obliterated. Fighting a war is itself a costly endeavour but—so the argument goes—not fighting and therefore submitting to the aggressor nation is (usually) more costly (Fotion 2007 p.10).

If an aggressor threatens to remove a people's autonomy, and all political response options are exhausted, this is regarded as a second legitimate reason to go to war. The autonomy (or freedom) argument is so important that it is considered even when it is indirectly threatened, for example, through the descent into chaos:

> If a nation's government is simply unable to govern effectively so that chaos results…If a government has lost control of its affairs so that chaos is everywhere…much suffering takes place amidst the chaos…chaos leads to a loss of people's autonomy (Fotion 2007 p.12).

Survival reasons related to food and water are considered a third just reason to mobilise or fight; although, as has long been documented in the literature, historical analysis reveals that humans vastly prefer establishing cooperative resource-sharing arrangements (De Stefano et al. 2012; Wolf 2007). Alternatively, as is occurring now in Sub-Saharan Africa, where food insecurity is already high and expected to worsen until 2025, instead of fighting, many simply endure hardship (Rosen et al. 2015). Food and water security issues can rarely be analysed in isolation; they are usually understood to be a consequence of multiple interspersed factors such as weak institutions, population growth, low development, poverty, violence, and crime (The World Bank 2011). Yet it is also well established that global warming will exacerbate food and water pressures for vulnerable humans and other species (Myers et al. 2017).

Actions that damage food supplies have sometimes been used as military tactics. Indeed, it was the standard practice of Ancient Greek armies to destroy enemy crops (Hanson 1998), and Agent Orange was used for the same effect in the Vietnam War. The hyperthreat will use this horrible tactic; eroding the capacity to grow food on land, attacking the availability of seafood and freshwater fish, and affecting the availability of fresh water (IPCC 2014).

Realists regard JWT as mere window-dressing and believe that, ultimately, most nations go to war to protect their interests (Fotion 2007 p.28). Even if this is so, the *protect their interests* argument could be applied to the hyperthreat. Regardless, JWT is important for establishing legitimacy, being granted resources, gaining the support of the populace, and gaining the commitment of individuals to fight.

To properly consider the *just cause* element of responding to the hyperthreat, the literature of climate ethics— from fields such as philosophy and theology—needs to be incorporated. Pope Francis's encyclical provides an exemplar because he links security, ethics, and climate change. He writes: 'peace, justice, and the preservation of creation are three absolutely interconnected themes, which cannot be separated' (Francis 2015 p.57)—a view which echoed across much of the environmental and human security literature discussed earlier.

Like Morton, Pope Francis is frank in describing the magnitude of the threat, as well as the secondary problem that humans 'do not grasp the gravity' of their situation (Francis 2015 p.65). Yet, when analysing the threat, Francis does not decentre the human; rather, he portrayed humans as the key agent—the warring party that is plundering and destroying sacred forms of life. He seeks to re-sensitise people to the true violence of their actions. For example, he highlights other theological deliberations about how the *do not kill* ethic might be applied in a globalised world—where depriving a group of resources necessary for life is an indirect form of killing (Francis 2015 p.59). Thus, Francis's view introduces a third version of culpability–*–unconscious culpability*—which also relates to the separation factor problem between the hyperthreat and victims, as discussed earlier.

If humans are the destructive force——or threat—then, Francis concludes, the driving force behind this enemy is cultural:

> We still lack the culture needed to confront this crisis. We lack the leadership capable of striking out on new paths (Francis 2015 p.34).

Pertinent to Morton's use of object-oriented ontology,, Francis contemplates subject-object-type human relationships with other forms of life and matter, finding that whereas it was once in tune and respectful, it is now extractive. Humans do not perceive that they and other life forms are intricately connected, he explains, therefore they do not equate war on nature with war upon themselves. Francis posits that it is humans' flawed culture, or world view, which allows indifference to suffering and ignorance of interconnectedness or 'the delicate equilibria existing between the creatures of this world' (Francis 2015 p.43). Akin to Beck's notion of 'zombie institutions,' Francis argues that current institutions and regulatory processes fail because they do not have the capacity to consider future generations.

Overall, the rationale behind JWT—the broader reasons that might justify large-scale mobilisation against an aggressor—may be applicable to the hyperthreat of climate and environmental change. They are also important because they relate to von Clausewitz's *moral forces* and therefore impact *fighting spirit*—or motivation to act.

To summarise JWT and hyperthreat deliberations, there are three *good reasons* to challenge the hyperthreat. Firstly, on destruction, it will kill and harm as a warlike aggressor might. There will be the obvious direct killing that occurs through extreme weather events such as Hurricane Katrina, and the extreme heatwaves or mega droughts and associated damage to property and infrastructure. More significant are indirect slow violence impacts; the loss of livelihoods, and the reduction in ability to grow food, drink clean water, or breathe fresh air. Taking a non-human-centric viewpoint, the outcome of wide-scale ecocide must be considered.

Secondly, autonomy is attacked from multiple directions. If allowed to reach its full fighting strength, the hyperthreat threatens chaos, which in turn affects autonomy. The autonomy of future generations will suffer attrition warfare, through which, one by one, choices disappear. These choices may disappear without people having known they once existed. Thirdly, survival reasons are self-evident; food and water availability are vulnerable and in some cases unable to be fortified against hyperthreat methods of attack. Habitability of land and dwellings for humans and other species is also at stake.

The JWT analytic lens reveals something else—people are removed from and desensitised to the damage they inflict, thereby becoming unconscious agents of the hyperthreat. Meanwhile, those who are conscious saboteurs are spared the ethical safety check of being able to see or hear the impacts of their actions upon their victims.

5.2 *Hyperthreats Defy Existing Nation-State Security Logic*

Broadly, nation-states seek to create peaceful, prosperous communities with a high quality of life for citizens. To achieve these aspirations, citizens fund defence forces to protect their nation-states from aggressors who may threaten these aims. A strong state can afford a strong military to protect it and allow it to flourish. Climate defies this logic. The hyperthreat of climate and environmental change erodes the nation-state's strength and resources from within, thereby also affecting the nation's capacity to sustain military forces. The hyperthreat enemy bypasses the nation-state's ability to protect itself because it pretends it is a mere hazard. Like a child with a backpack containing a deadly biological weapon who is ushered through the security checkpoint regardless, the hyperthreat evades normal threat analysis and response. Extant notions of threat act as blinkers, preventing security planners from properly seeing or critiquing the hyperthreat.

5.3 *Summary Description of the Hyperthreat*

Drawing together various insights, a working definition of the hyperthreat is offered as follows. The hyperthreat has warlike destructive capabilities that are so diffuse that it is hard to see the enormity of the destruction coherently nor who is responsible for its hostile actions. It defies existing human thought and institutional constructs. It is powered and energised by three key enablers; its invisibility, its ability to evade all existing human threat-response mechanisms, and by human hesitancy—the slower humans are to act, the stronger it becomes.

5.4 How to Talk about the Hyperthreat

Deep framing research argues that introducing new thought needs to begin with precognitive communication processes; the use of metaphors, imagery, sensory, and affective signals. Stories evoke and integrate these dimensions of knowing. Given the breadth of research on climate change communication that is critical of so-called rational, facts-based, and linear communication methods, describing a hyper-threat must be far more than listing likely numbers of deaths or types of possibilities of harm—an approach which has already dominated climate discourse.

Returning briefly to Morton, he argues that without the aesthetic, the full understanding of a threat remains limited. Referring to his own experience of listening to JLiat's recording of hydrogen bomb tests in the Pacific, he wrote, 'hearing it, rather than seeing it…restores to the aesthetic dimension a trauma and a pain that we edit out at our peril' (Morton 2013 p.193).

Cowering to the dismissive frame of *alarmism* when conveying the truth about the scale of threat involved is not ethical or honest. Climate and environmental change have hyper-dimensions that need to be conveyed in a way that allows people to comprehend deeply the true scale of the threat they face. For example, greenhouse gas emissions could be conceived as being like destructive missiles that, although launched gradually from mid-twentieth Century and now via daily volleys, will not land or have their most destructive effects for approximately 30 years. Australia's Great Barrier Reef can be recognised as an early casualty—one of the first to fall at the hand of the hyperthreat (Tarte et al. 2017).

These types of metaphors, and the utilisation of skilled creative talent to convey the feeling of the threat, (far better than can be achieved here) is a potential tactic to help defeat the hyperthreat—because it counters the hyperthreat's invisibility (Boulton 2016b).

5.5 Limitations and Policy Implications of the Hyperthreat Framing

The greatest concern with the hyperthreat framing is that in defining a new type of enemy, an adversarial perspective is adopted, which may destroy the possibility of constructive dialogue and cooperation with elements of, for example, the fossil fuel industry. In addition, it could instil fear in people, which could lead to severe or ill-considered responses. However, it is argued here that the identification of a threat with hostile intent does not imply that a militaristic response is required. It suffices to identify now that if part of the hyperthreat construction is cultural, with an ability to defy existing governance norms, this might logically suggest where the strongest solutions may be found.

The aim of this chapter was to define the hyperthreat; crafting a hyper-response requires more extensive work. However, at this early stage, some immediate policy implications of a hyperthreat frame can be suggested. Primarily, before pragmatic aspects of the problem are tackled, policy needs to re-engage with *moral forces*—the philosophical, ethical, and affective dimensions of the problem. Philosophy must re-enter public discourse to help reconsider notions like 'security' or the protection of the populace anew in a climate-impacted era. Whose responsibility is human security, for example? Secondly, it needs to be accepted that security strategy in the face of climate and environmental change must involve a wide array of disciplinary perspectives and the cross-pollination of expertise across old boundaries.

6 Conclusion

> Non-human beings are responsible for the next moment of human history and thinking (Morton 2013 p.201).

Research into the muted and inadequate response to global warming points to the criticality of prevailing grand narratives or deep frames that shape the way people and their institutions think. Timothy Morton's hyperobject framing opens a new way of thinking, yet it may not capture the problem accurately enough to be the definitive guiding compass.

The 'hyperthreat' frame seeks to develop Morton's conception but also explore what happens when the hyperobject notion is placed within the climate-security debate. It agrees with Morton and Beck that a solution may lie outside current zombie-like, siloed concepts, and institutional and governance arrangements. Accordingly, it departs from both the traditional security sector's pre-climate notions of what a threat is, and the laboured climate-security discourse that focuses mostly upon the different ways climate may or may not hurt people or contribute to the likelihood of violent conflict. To avoid the trap of pre-climate era conceptual straitjackets, the hyperthreat notion returns to fundamental questions and the starting point of any military strategy: a threat analysis. What sort of threat is it? How does it move and appear? How will it destroy? Who is behind it?

In so doing, the hyperthreat notion brings into hard, sharp relief notions that many hesitate to acknowledge. There is a threat—a different type of threat. The hyperthreat will act like an aggressive enemy by damaging people's livelihoods, property, safety, health, autonomy, and in some cases, ability to survive. Its primary victims—nature and future generations—are mostly voiceless; their cries are not heard. The threat's main method of operating involves infusion; it hides among the folds of modern-era economic, justice, and governance systems, thereby evading detection and scrutiny. It bypasses the traditional security sector's watch by virtue of mere word definitions, appearing as a hazard; not a threat with a conscious, hostile intent.

An initial understanding of the hyperthreat's dimensions also brings into fresh sight some human tools and strengths that may hitherto have been overlooked, such as courage and knowledge about how to mobilise around a common threat. Humans have the skills to protect vulnerability, and may be able to recover the motivation to do so. Moreover, if humans have been blinded to the scale of the problem, their full potential to respond to the hyperthreat may not yet have come into view either. Removing modern-era conceptual blinkers may reveal astounding horizons and possibilities.

References

Adger, W. N., Pulhin, J. M., Barnett, J., Dabelko, G. D., Hovelsrud, G. K., Levy, M., Oswald Spring, U., & Vogel, C. H. (2014). Human security. In C. B. Field, V. R. Barros, D. J. Dokken, K. J. Mach, M. D. Mastrandrea, T. E. Bilir, M. Chatterjee, K. L. Ebi, Y. O. Estrada, R. C. Genova, B. Girma, E. S. Kissel, A. N. Levy, S. MacCracken, P. R. Mastrandrea, & L. L. White (Eds.), *Climate change 2014: Impacts, adaptation, and vulnerability. Part a: Global and Sectoral aspects. Contribution of working group II to the fifth assessment report of the intergovernmental panel on climate change* (pp. 755–791). Cambridge, UK/New York, USA: Cambridge University Press.

Allen, J. R., & Husain, A. (2017). On Hyperwar. *Proceedings Magazine* (Vol. 143). Annapolis: U.S. Naval Institute.

Arendt, H. (1963). *Eichmann in Jerusalem: A report on the banality of evil*. New York: Penguin.

Barnett, J. (2001). *Security and climate change*. vol Working Paper 7. Tyndall Centre for Climate Change Research, London, UK.

Barnett, J. (2007). Environmental security and peace. *Journal of Human Security, 3*(1), 4.

Barnett, J. (2008). Peace and development: Towards a new synthesis. *Journal of Peace Research, 45*(1), 75–89. https://doi.org/10.1177/0022343307084924.

Barnett, J. (2009). The prize of peace (is eternal vigilance): A cautionary editorial essay on climate geopolitics. *Climatic Change, 96*(1–2), 1–6. https://doi.org/10.1007/s10584-009-9591-5.

Bennett, J. (2009). *Vibrant matter: A political ecology of things*. Duke University Press.

Blaikie P, Cannon T, Davis I, Wisner B. (2014). *At risk: natural hazards, people's vulnerability and disasters*. Routledge.

Bliuc, A.-M., McGarty, C., Thomas, E. F., Lala, G., Berndsen, M., & Misajon, R. (2015). Public division about climate change rooted in conflicting socio-political identities. *Nature Climate Change, 5*(3), 226–229. https://doi.org/10.1038/nclimate2507. http://www.nature.com/nclimate/journal/v5/n3/abs/nclimate2507.html#supplementary-information.

Bodansky, D. (2010). The Copenhagen climate change conference: A postmortem. *American Journal of International Law, 104*(2), 230–240.

Boulton, E. (2016a). Climate change as a 'hyperobject': A critical review of Timothy Morton's reframing narrative. *Wiley Interdisciplinary Reviews: Climate Change*, *7*(5),772–785. https://doi.org/10.1002/wcc.410.

Boulton E. (2016b). It's time for a new age of Enlightenment: why climate change needs 60,000 artists to tell its story. *The Conversation*, June 2016.

Boyne, R. (2001). Cosmopolis and risk: A conversation with Ulrich beck. *Theory, Culture & Society, 18*(4), 47–63.

Brauch, H. G. (2002). Climate change, environmental stress and conflict. *Climate change and conflict*, 9–112.

Brauch, H. G. (2008). *From a security towards a survival dilemma. Globalization and environmental challenges: Reconceptualizing security in the 21 st century* (pp. 537–552).

Brown, B. (2012). *Daring greatly: How the courage to be vulnerable transforms the way we live, love, parent, and lead*. Penguin.

Brown, L. R. (2008). *Plan B 3.0: Mobilizing to save civilization* (Vol. 384). New York: WW Norton.
Brunstetter, D., & Braun, M. (2011). The implications of drones on the just war tradition. *Ethics & International Affairs, 25*(03), 337–358.
Buhaug, H., Nordkvelle, J., Bernauer, T., Böhmelt, T., Brzoska, M., Busby, J. W., Ciccone, A., Fjelde, H., Gartzke, E., & Gleditsch, N. P. (2014). One effect to rule them all? A comment on climate and conflict. *Climatic Change, 127*(3–4), 391–397.
Carson R. (1962). Silent Spring.
Chaliand, G. (1994). *The art of war in world history: From antiquity to the nuclear age*. Univ of California Press.
Chamayou, G. (2015). *Drone theory*. UK: Penguin.
Cheal, D. (2015). *The gift economy* (Vol. 4). Routledge.
Clapper, J. (2016). *Statement for the record worldwide threat assessment of the US intelligence community senate armed services committee. Office of the Director of National Intelligence*. USA: Washington.
Clausewitz VC. (1832). *On war* (trans: Colonel J.J. Graham). Penguin Books, London.
CNA MAB. (2014). *National Security and the accelerating risks of climate change*. Alexandria: CNA Military Advisory Board.
Connell, R. W., & Messerschmidt, J. W. (2005). Hegemonic masculinity rethinking the concept. *Gender & Society, 19*(6), 829–859.
Courtenay, W. H. (2000). Constructions of masculinity and their influence on men's well-being: A theory of gender and health. *Social Science & Medicine, 50*(10), 1385–1401.
Crutzen, P. J. (2006). The "anthropocene". In *Earth system science in the Anthropocene* (pp. 13–18). Springer.
Dabelko, G. D. (2009). Planning for climate change: The security community's precautionary principle. *Climatic Change, 96*(1), 13–21.
Daggett, C. (2014). Morton, Timothy 2013 Hyperobjects: Philosophy and ecology after the end of the world, reviewed by Cara Daggett. In *Environment and planning D: Society and space*.
Day D AM Stephen (Major General). (2015). *Thoughts on generalship: Lessons from two wars*. Australian Army, Canberra.
De Stefano, L., Duncan, J., Dinar, S., Stahl, K., Strzepek, K. M., & Wolf, A. T. (2012). Climate change and the institutional resilience of international river basins. *Journal of Peace Research, 49*(1), 193–209.
DeMarco, P. M. (2017). Rachel Carson's environmental ethic–a guide for global systems decision making. *Journal of Cleaner Production, 140*, 127–133.
Di Leo, J. R. (2013). Can theory save the planet?: Critical climate change and the limits of theory. *Symploke, 21*(1), 27–36.
Diamond, J. (2005). *Collapse: How societies choose to fail or succeed*. Penguin.
Dubash, N. K. (2009). Copenhagen: Climate of mistrust. *Economic and Political Weekly*, 8–11.
Dunlap, R. E., & McCright, A. M. (2011). Organized climate change denial. In *The Oxford handbook of climate change and society* (pp. 144–160).
Dupont A, Pearman G (2006) Heating up the planet. Paper 12.
Floyd, R. (2008). The environmental security debate and its significance for climate change. *The International Spectator, 43*(3), 51–65.
Fotion, N. (2007). *War & Ethics - a new just war theory*. London: Continuum International Publishing Group.
Francis, P. (2015). *Encyclical letter: Laudato Si' of the holy father Francis on Care for our Common Home*. Vatican: The Catholic Church.
Galschiøt, J. (1991). *Cocoon shield 1* (Vol. Over 2 metres). Jens: Galschiøt.
Galtung, J. (1969). Violence, peace, and peace research. *Journal of Peace Research, 6*(3), 167–191.
Gilding, P. (2011). *The great disruption: Why the climate crisis will bring on the end of shopping and the birth of a new world*. Bloomsbury Publishing USA.
Gleditsch, N. P., & Nordås, R. (2014). Conflicting messages? The IPCC on conflict and human security. *Political Geography, 43*, 82–90.

Hamilton, C. (2007). *Scorcher: The dirty politics of climate change*. Black.
Hannam P. (2015). Australia's military stuck in the 'wilderness' over climate change former ADF chief says, *The Sydney Morning Herald*, 24 Oct 2015.
Hansen, J., Kharecha, P., Sato, M., Masson-Delmotte, V., Ackerman, F., Beerling, D. J., Hearty, P. J., Hoegh-Guldberg, O., Hsu, S.-L., & Parmesan, C. (2013). Assessing "dangerous climate change": Required reduction of carbon emissions to protect young people, future generations and nature. *PLoS One, 8*(12), e81648.
Hanson, V. D. (1998). *Warfare and agriculture in classical Greece* (Vol. 40). Univ of California Press.
Heyward, M. (2007). Equity and international climate change negotiations: A matter of perspective. *Climate Policy, 7*(6), 518–534.
Hoggan, J., & Littlemore, R. (2009). *Climate cover-up: The crusade to deny global warming*. Greystone Books Ltd.
Holmes, S. (2012). Wither Ecocriticism in the era of the Hyperobject? A review of Timothy Morton's ecology without nature and the ecological thought. *Journal of Ecocriticism, 4*(1), 57–61.
Hsiang, S. M., & Burke, M. (2014). Climate, conflict, and social stability: What does the evidence say? *Climatic Change, 123*(1), 39–55.
Hulme, M. (2007). Climate conflict, *New Scientist, 196*(2629), 26.
Hyde, L. (2007). *The gift: How the creative spirit transforms the world*. Canongate Books.
IPCC. (2012). Managing the risks of extreme events and disasters to advance climate change adaptation. Special report of the Intergovernmental Panel on Climate Change.
IPCC. (2014). Climate Change 2014: Synthesis Report. Contribution of Working Groups I, II and III to the Fifth Assessment Report of the Intergovernmental Panel on Climate Change. IPCC, Geneva, Switzerland.
Kahan DM, Wittlin M, Peters E, Slovic P, Ouellette LL, Braman D, Mandel GN. (2011). The tragedy of the risk-perception commons: Culture conflict, rationality conflict, and climate change (trans: Law BSo). *Temple University legal studies research paper*, vol 26. Temple University, Philadelphia.
Kemp L. (2017). Better out than in. *Nature Climate Change, 7*(7), 458. doi: https://doi.org/10.1038/nclimate3309.
Klein N (2014) *This changes everything: capitalism vs. the climate*. New York: Simon and Schuster.
Klinsky, S., Waskow, D., Northrop, E., & Bevins, W. (2017). Operationalizing equity and supporting ambition: Identifying a more robust approach to 'respective capabilities'. *Climate and Development, 9*(4), 287–297.
Lakoff, G. (2010). Why it matters how we frame the environment. *Environmental Communication, 4*(1), 70–81. https://doi.org/10.1080/17524030903529749.
Lazarus, R. J. (2009). Super wicked problems and climate change: Restraining the present to liberate the future. *Cornell Law Review, 94*(5), 1153–1233.
Leonardi, E. (2014). Global warming: Between history and history. *Reviews in Cultural Theory, 5*(1), 71–75.
Lewandowsky, S., Oreskes, N., Risbey, J. S., Newell, B. R., & Smithson, M. (2015). Seepage: Climate change denial and its effect on the scientific community. *Global Environmental Change, 33*, 1–13.
Loewenstein, G. (2010). Insufficient emotion: Soul-searching by a former indicter of strong emotions. *Emotion Review, 2*(3), 234–239.
Loewenstein G, Schwartz D. (2010). Nothing to Fear but a Lack of Fear: Climate Change and the Fear Deficit. Paper presented at the G8/G20 Summit 2010, Ontario,Toronto,
Lovelock, J. E. (2009). *The vanishing face of Gaia: The final warning*. PublicAffairs.
Mathews, J. T. (1989). Redefining security. *Foreign Affairs, 68*(2), 162–177. https://doi.org/10.2307/20043906.
McGoey, L. (2012). The logic of strategic ignorance. *The British Journal of Sociology, 63*(3), 533–576.

Morton, T. (2013). *Hyperobjects philosophy and ecology after the end of the world*. Minneapolis: University of Minnesota Press.
Myers, S. S., Smith, M. R., Guth, S., Golden, C. D., Vaitla, B., Mueller, N. D., Dangour, A. D., & Huybers, P. (2017). Climate change and global food systems: Potential impacts on food security and undernutrition. *Annual Review of Public Health, 38*, 259–277. https://doi.org/10.1146/annurev-publhealth-031816-044356
Nielsen, E. B. (2017). Climate Crisis Made Manifest. In D. G. Ross (Ed.), *Topic-driven environmental rhetoric*. New York: Routledge.
Nisbet, M. C. (2009). Communicating climate change: Why frames matter for public engagement. *Environment: Science and Policy for Sustainable Development, 51*(2), 12–23.
Nixon, R. (2011). *Slow violence and the environmentalism of the poor*. Harvard University Press.
Nussbaum, M. C. (2003). *Upheavals of thought: The intelligence of emotions*. New York: Cambridge University Press.
Oppenheimer, M., Campos, M., Warren, R., Birkmann, J., Luber, G., O'Neill B., & Takahashi K. (2014). Emergent risks and key vulnerabilities. In: C. B. Field, V. R. Barros, D. J. Dokken, K. J. Mach, M. D. Mastrandrea, T. E. Bilir, M. Chatterjee, K. L. Ebi, Y. O. Estrada, R. C. Genova, B. Girma, E. S. Kissel, A. N. Levy, S. MacCracken, P. R. Mastrandrea, & L. L. White (eds.), *Climate change 2014: Impacts, adaptation, and vulnerability. Part A: Global and sectoral aspects. Contribution of working group II to the fifth assessment report of the intergovernmental panel on climate change*. United Kingdom/New York.
Oreskes, N., & Conway, E. M. (2010). *Merchants of doubt*. New York: Bloomsbury.
Palsson, G., Szerszynski, B., Sörlin, S., Marks, J., Avril, B., Crumley, C., Hackmann, H., Holm, P., Ingram, J., & Kirman, A. (2013). Reconceptualizing the 'Anthropos' in the Anthropocene: Integrating the social sciences and humanities in global environmental change research. *Environmental Science & Policy, 28*, 3–13.
Parmesan, C., & Yohe, G. (2003). A globally coherent fingerprint of climate change impacts across natural systems. *Nature, 421*(6918), 37–42.
Penny, C. K. (2007). Greening the security council: Climate change as an emerging "threat to international peace and security". *International Environmental Agreements: Politics, Law and Economics, 7*(1), 35–71.
Rockström, J., Steffen, W., Noone, K., Persson, A., Chapin, F. S., Lambin, E. F., Lenton, T. M., Scheffer, M., Folke, C., Schellnhuber, H. J., Nykvist, B., de Wit, C. A., Hughes, T., van der Leeuw, S., Rodhe, H., Sorlin, S., Snyder, P. K., Costanza, R., Svedin, U., Falkenmark, M., Karlberg, L., Corell, R. W., Fabry, V. J., Hansen, J., Walker, B., Liverman, D., Richardson, K., Crutzen, P., & Foley, J. A. (2009). A safe operating space for humanity. *Nature, 461*(7263), 472–475.
Rosen, S., Meade, B., & Murray, A. (2015). *International food security assessment, 2015–2025* (Vol. GFA-26). Washington: US Department of Agriculture, Economic Research Service.
Roy, A. (2014). *Capitalism a ghost story*. London: Verso.
Rüttinger L, Smith D, Stang G, Tänzler D, Vivekananda J. (2015). A new climate for peace taking action on climate and fragility risks (trans: Woodrow Wilson International Center for Scholars, Studies EUIfS). G7, Paris.
Salehyan, I. (2014). *Climate change and conflict: Making sense of disparate findings*. Elsevier.
Schwartz, P. D. R. (2003). *An abrupt climate change scenario and its implications for United States National Security*. California: Jet Propulsion Laboratory Pasadena.
Selby J, Dahi OS, Fröhlich C, Hulme M (2017) Climate change and the Syrian civil war revisited. *Political Geography, 60*, 232-244
Spring, Ú. O. (2009). A HUGE gender security approach: Towards human, gender, and environmental security. In *Facing global environmental change* (pp. 1157–1181). Springer.
Steffen, W., Crutzen, P. J., & McNeill, J. R. (2007). The Anthropocene: Are humans now overwhelming the great forces of nature. *Ambio: A Journal of the Human Environment, 36*(8), 614–621.
Stott, P. A., Christidis, N., Otto, F. E. L., Sun, Y., Vanderlinden, J.-P., van Oldenborgh, G. J., Vautard, R., von Storch, H., Walton, P., Yiou, P., & Zwiers, F. W. (2016). Attribution of extreme

weather and climate-related events. *Wiley Interdisciplinary Reviews: Climate Change, 7*(1), 23–41. https://doi.org/10.1002/wcc.380.
Swyngedouw, E. (2010). Apocalypse forever? *Theory, Culture & Society, 27*(2–3), 213–232.
Tabuchi H. (2017). 'Vulnerable Voices' lash out as companies sway climate talks. *The New York Times*, 16 May 2017.
Tänzler, D., Maas, A., & Carius, A. (2010). Climate change adaptation and peace. *Wiley Interdisciplinary Reviews: Climate Change, 1*(5), 741–750.
Tarte, D., Hart, B., Hughes, T., & Hussey, K. (2017). *Reef 2050 long-term sustainability plan progress on implementation review*. Sydney: Great Barrier Reef Independent Review Group.
The World Bank. (2011). *World development report: Conflict, security, and development. International bank for reconstruction and development*. Washington: The World Bank.
Thomas, M. (2013). The securitisation of climate change: A military perspective. *Australian Defence Force Journal, 192*, 7.
Tooby, J., & Cosmides, L. (2008). The evolutionary psychology of the emotions and their relationship to internal regulatory variables. In M. Lewis, J. M. Haviland-Jones, & L. F. Barrett (Eds.), *Handbook of emotions* (3rd ed., pp. 114–137). New York: Guilford Press.
UNEP. (1992). Rio declaration on environment and development. United Nations. http://www.unep.org/Documents.Multilingual/Default.asp?documentid=78&articleid=1163. Accessed 06 March 2014.
UNFCCC. (2017). Joint civil society submission regarding the Arrangements for Intergovernmental Meetings in-session workshop on opportunities to further enhance the effective engagement of Non-Party Stakeholders at UNFCCC SBI 46 (May 2017). Secretariat of the United Nations Framework Convention on Climate Change. https://unfccc.int/files/parties_and_observers/application/pdf/855final.pdf. Accessed 19 May 2017.
US DOD. (2011). *Joint publication 5–0 joint operation planning*. Washington: Joint Chiefs of Staff.
Wolf, A. T. (2007). Shared waters: Conflict and cooperation. *Annual Review of Environment and Resources, 32*, 241–269.
Woodbridge, R. M. (2004). *The next world war: Tribes, cities, nations and ecological decline*. Toronto: University of Toronto Press.

Chapter 6
Religion and Australia's Near Region

Dan Cassidy

1 Introduction

Australia, like most Western countries, makes a clear separation between religion and state. The Westminster system of government in Australia encapsulates this separation, with the three branches of government: legislative, executive and judicial. In this social-political system, there is no formally recognised government religion or church, and no formally recognised mechanism for one or more religious bodies to shape or significantly influence these three pillars. In fact, Australia's foundations were never based on significant religious conviction or event. In his address to the Adelaide sessions of the Australasian Constitutional Convention Debates, future inaugural Prime Minister Edmund Barton enshrined the ideal of government rule over religion (although it has never been absent from the nation's discourse and character):

> "The whole mode of government, the whole province of the State, is secular... The whole duty is to render unto Caesar the things that are Caesar's, and unto God the things that are God's. That is the line of division maintained in every State in which there is not a predominant church government which dictates to all civil institutions... The best plan which can be adopted as to a proposal of this kind, which is so likely to create dissension foreign to the objects of any church, or any Christian community, is that secular expressions should be left to secular matters while prayer should be left to its proper place." (Barton 1897)

This separation of religion and state has meant that strategic and military planners in practice have not been required to develop mechanisms for controlling or mitigating religious ideologies or worldviews which could unduly subvert or threaten the independence of the state. It comes in stark contrast to Australia's near

D. Cassidy (✉)
2nd Cavalry Regiment, Royal Australian Army Chaplains Department, 3rd Brigade, Townsville, QLD, Australia
e-mail: Daniel.cassidy@defence.gov.au

S. Pearson et al. (eds.), *Australian Contributions to Strategic and Military Geography*, Advances in Military Geosciences,
https://doi.org/10.1007/978-3-319-73408-8_6

region or Primary Operational Environment (POE). For the purpose of this discourse, Australia's POE is taken to be Southeast Asia and the Southwest Pacific, which is enveloped with nation states that are heavily influenced by the fusion of religion and state.

This chapter is designed as an introduction, highlighting how important religion is to strategic and military geography. In this context, it is critical for all those policymakers and planners, and those who are deployed in our POE, to acknowledge and develop a keen awareness of the similarities and differences between the role that religion plays in these systems as distinct from our own.

1.1 Perspective

Different socio-political groups have certain perspectives by which they see the world. All people have a deep need for meaning and it is not enough for people to simply view the world. They seek to understand what they see and experience in order to fit these experiences into their perspective of the world—their *Weltanschauung* or worldview (Olthuis 1989). Worldviews have been described as the set of assumptions or pre-suppositions people use to make sense of life. These assumptions are held at a very fundamental level and provide a comprehensive framework for one's basic beliefs about life and ultimately behaviour.

It is important to note that everybody has a worldview of some kind; a worldview in some form is fundamental to existence. The theologian Francis Schaefer (1982) expressed it this way:

> "In this sense, all people are philosophers, for all people have a world-view. This is as true of the man digging a ditch as it is of the philosopher in the university."

1.2 Worldview and Religion

The Enlightenment, and later Postmodernism, have brought with them a secular worldview which has largely subjugated any reference to a religious belief or worldview. As a result, the modern West cannot understand why those with a religious worldview act differently and are ultimately willing to die for their cause. This historical exclusion of religion has come about as an outgrowth of secularism and the separation of religion and state in liberal democratic societies. By making this clear demarcation between religion and state, Western governments, including Australia, fail to fully appreciate how religious belief shapes the perceptions and political aspirations of others who do not similarly separate the two.

For example, Islam plays an integral part in the Malay worldview (Verma 2002). In PNG, outside regional capitals all communities belong to and are to a certain extent administered by a denomination of the church, with up to 90% of health and education being managed and delivered by the churches; they are the service delivery arm of government.

People from different worldviews or religions think about the world in different ways; they will also behave in different ways.

> "Religion can create cultural instability, but in many cases, it can alleviate problems in a society and be a unifying force... The effect of a religion has to be understood within that religion and within that society" (United States Army 2014).

To understand why they do what they do, Western military planners need to try and see the world the way they do (see, for example, Durward and Marsden 2016). Therefore, it is argued that understanding the place of religion, especially within Australia's near region, is one of the most important aspects in meeting the challenges of contemporary conflict.

1.3 Practice

Military practitioners struggle to address the religious dimension to conflict. Typically, they have not included "a sophisticated understanding of the larger religious" and socio-political values that influence actors in a conflict (Johnston and Sampson 1994). They have failed to appreciate the fusion of a spiritual worldview and politics or what might be defined as the fusion of 'religion and state.' This often limits their focus to the traditional military appreciation process (a logical decision-making process that is meant to analyse all the relevant factors in a situation and coordinates a headquarters to develop an appropriate course of action) and the application of hard kinetic effects. The absence of a much fuller appreciation of religion's place in conflict and politics prevents governments from providing a clear end-state or grand strategy to solve the problem. It also reduces prevention and post-conflict possibilities. The result is a simplistic 'whack a mole' approach to perceived or realised threats and the expenditure of limited treasury funds.

In the most recent literature on this topic, religion is given little attention; except to provide a secular category to name the adversary that is being fought. This in turn encourages analysts to think of religious beliefs, like Islam, as a monolithic entity and treat it accordingly (Kilcullen 2005; Barno 2006; Arif 2013). The Australian military has generally failed to see the inherent value in understanding these elements of power and influence, especially within its near region. Michael Evans points out that:

> "An insufficient attempt has been made to undertake a careful analysis of the differences between elements of power (the combination of liberal democracy, political economy and culture) and the instruments of power (the combination of soft and hard power, interagency and military force) necessary to secure national interests" (Evans 2008).

Finally, in 2009 and again in 2017, Major General Roger Noble (Australian Army) and Lieutenant General Gary Volesky (U.S. Army) have argued for strategists to focus on people, culture and society.

> "Operations in Iraq in 2016 once again confirmed the basic observation that wars are fought by people for human ends and purposes … ultimately people (on both sides) decide whether they have won or lost, not platforms or systems." (Noble and Volesky 2017).

The human domain should be an indispensable and enduring strand in all strategic calculations.

> "The realm of strategic theory, long dominated by political scientists, Cold War warriors, game theorists and international relations experts, has ignored, or been unaware of, the insights and illumination that can be gained from paying close attention to the work of the 'soft sciences'." (Noble 2009).

2 Australia's Primary Operational Environment (POE)

Across the region, the role of religion in society, politics and national identity has grown. The twentieth century was a time of dramatic shifts in religious expression. In light of the even-broader values, including secularism, pluralism and liberalism, that globalization imposes onto the world, many regions, countries and ethnic groups react by reasserting a more traditional or fundamental expression of their faith. This goes for much of the Muslim world, the Hindu world, many contexts of traditional ethnic religions and even Christianity. Analysis of different actors' involvement, relationships, religious beliefs, motives, perceptions, interested and desired outcomes is integral to understanding the operating environment, especially in Asia and the Pacific. The patterns of religious belief can be of predictive value and of particular relevance to military analysis and planning, and assist in identifying possible individual group behaviour. Such analyses provide a valuable point-in-time assessment, however, greater insights can be gained by looking at the relative changes between the major religious groups over time. Table 6.1 shows the sizes and projected growth of major religious groups in Asia-Pacific, 2010–2050. These projected shifts in major religious groups arguably can also assist military strategists consider projected medium and longer-term scenarios.

2.1 Asia – Religion

Many Asian states actively promote state-based religions and enforce limitations on individual and organisational evangelism. Despite the lack of religious freedom and the reality of widespread religious persecution, when compared to other continents, religious belief is most prominent in Asia. There is amazing religious diversity as well as significant differences in both religious tolerance and the role religion plays in society. This stretches from diverse multi-religious societies such as those in India to virtually mono-religious nations such as the Yemen; from secularism or even state-endorsed atheism as in the North Korea to theocracies like the Islamic

Table 6.1 Size and projected growth of major religious groups in Asia-Pacific, 2010–2050. Population estimates are rounded to the nearest 10,000. Percentages are calculated from unrounded numbers. Figures may not add to 100% because of rounding. Adapted from: The Future of World Religions: Population Growth Projections 2010–2050. Adapted from Pew Research Center (2015)

	2010 estimated population	% in 2010	2050 projected population	% in 2050	Population growth 2010–2050	Increase 2010–2050	Compound annual growth rate (%)
Hindus	1,024,630,000	25.3	1,369,600,000	27.7	344,970,000	33.7	0.7
Muslims	986,420,000	24.3	1,457,720,000	29.5	471,290,000	47.8	1.0
Unaffiliated	858,490,000	21.2	837,790,000	17.0	−20,700,000	−2.4	0.1
Buddhists	481,480,000	11.9	475,840,000	9.6	−5,640,000	−1.2	0.0
Folk religions	364,690,000	9.0	366,860,000	7.4	2,170,000	0.6	0.0
Christians	287,100,000	7.1	381,200,000	7.7	94,100,000	32.8	0.7
Other religions	51,920,000	1.3	48,650,000	1.0	−3,280,000	−6.3	0.2
Jews	200,000	<0.1	240,000	<0.1	40,000	21.2	0.5
Regional total	4,054,930,000	100.0	4,937,900,000	100.0	882,950,000	21.8	0.6

Republic of Iran, and the Kingdom of Saudi Arabia with its unitary Islamic absolute monarchy.

Asia is home to three other major world religions apart from Christianity—Islam, Hinduism and Buddhism. It is also the heartland of several other significant religions—Sikhism, Jainism, Daoism, Confucianism, Baha'i, and Judaism. Apart from the latter, Asia contains the greatest concentration of the adherents of all of these groups. In Australia's near region, the strongly Muslim character of Malaysia and Indonesia is readily apparent.

2.2 *Pacific – Religion*

The Pacific comprises one continent (Australia), two large land masses (New Zealand and Papua New Guinea [PNG]) and 26 smaller island states and territories, but comprises a large Christian majority. For example, Christianity is the predominant belief system in PNG, with 96% self-identifying as Christian. The PNG Defence Organisation also places significant emphasis on its relationship with God. An entire chapter of their Defence White paper is dedicated to 'Developing Our Spiritual Strength' as a key dimension of capability:

> "At the heart of the PNGDO's existence is the spiritual welfare of its personnel and their families. In an ever-changing world of complex challenges that affect the family unit, the PNGDO must strengthen its relationship with God. The PNG Constitution fundamentally upholds our Christian values and on the 26th of August 2007, our nation signed a Covenant to remain true to God. We must bring the PNGDO close to God, remain true to the Covenant, and apply the Leadership example of Service that our Saviour Jesus Christ has bequeathed to us." (PNGDO 2013)

Another example is the Sovereign Democratic Republic of Fiji. Here much of the political machinery favours Christianity against minority religions. In fact, it is reputed that Fiji is tightly affiliated with Methodism, with the Methodist Church affirmed as being "the de facto state church in Fiji for 150 years as well as the faith of the majority of ethnic Fijians" (Mandryk 2010). In the 1996 Census, over 66% of ethnic Fijians were recorded as Methodist (Fiji Bureau of Statistics 2017).

In the Republic of Vanuatu the national motto is "Long God Yumi Stanap" ("In God We Stand"). Committed Christians played a major role in this nation attaining independence. The emphasis on their relationship with God is evidenced by this motto being emblazoned on their currency (shown in Fig. 6.1) and the continued role that Christians play in leading the country, not least the President himself. Former Presidents John Bani and Baldwin Lonsdale were clergymen, as is the current President, Tallis Obed Moses.

The national mottos of Samoa, Tonga, and Fiji also have a Christian character, being "Fa'avae i le Atua Sāmoa" ("Samoa is founded on God"), "Ko e 'Otua mo Tonga ko hoku tofi'a" ("God and Tonga are my inheritance"), and "Rerevaka na Kalou ka Doka na Tui" ("Fear God and honour the Queen").

Fig. 6.1 Vanuatu coin

3 A Practical Example

Travel to the Pacific Islands means seeing the indelible place that God has in the life of community. There is usually a church or temple in every village or main street, which is well attended on holy days. This is but one testament to the significance that religion and spirituality have to the people across the Pacific. Yet for decades, religion, spirituality and faith have been marginalised by well-intentioned foreign agencies in their development theory, policy and practice (Lunn 2009). The deep faith in the Pacific Islands is antithetical to the secular humanist push that is included in nearly all foreign state-funded development aid.

One such example is the case of climate change, where such agencies have sought to help the region's people adapt to the effects of future of climate change, often drawing up plans in a secular way. Patrick Nunn from the Australian Centre for Pacific Islands Research and Sustainability Research Centre, Australia, states that over three decades, most such interventions have failed (Betzold 2015), "proving neither effective nor sustainable" (Nunn 2017). He argues that the reason why "may in part lie in the side-lining of God."

In the past few years, representatives of these agencies have begun to acknowledge that modifying disaster risk management and developmental programs to include spirituality rather than purely using secular terms is critical to success. Secular messages fail to connect with the spiritually orientated communities and, in some instances, can be met with indifference or even hostility if they clash with the communities' spiritual outlook. The messages that resonate best in many poorer countries, including the Pacific Islands, are those that engage with people's spiritual beliefs and which are channelled through religious leaders (Nunn 2017). These experiences suggest that military forces engaged in peacekeeping and humanitarian relief operations also need to respect and engage with local spiritual beliefs and religious leadership in order to succeed in their mission.

4 Challenges Ahead

Southeast Asia and the Southwest Pacific form Australia's near region and, in this region, religion is part of the context within which military operations and routine relationships are conducted (Wunderle 2006). The problem today is that Australia's foremost strategic thinkers and policy advisors operate only from a secular worldview, which underestimates the significance of religion to the region.

Campaigns in Afghanistan and Iraq should have highlighted the importance of religion in strategy. There is a requirement to address not only the 'guns', but also the ideas behind the guns, the worldview that fuses the spiritual and the political, in order to bring about conflict transformation. When religion is not incorporated into campaign planning, the Australian Defence Force (ADF) is left vulnerable. Militaries like the ADF can find, fix and destroy an adversary, but are ill equipped or designed to attack the conditions that make religiously motivated conflict viable.

Australia must, therefore, engage holistically in the world with a realism that respects the political and spiritual power of religion (Seiple 2007). Ignoring the significance of religion to Australia's near region increases the risk of mission or campaign failure, and creates barriers to successful interaction. Leaders from across government must approach things differently in order to combat this threat. This includes military leaders who must be "agile, well-informed, culturally astute," and who at least understand that religion should not be "the missing dimension in mission planning" (Bedsole 2006).

> "Some may realize that while religion might be a negative catalyst to conflict, faith can serve as a positive component of sustainable solutions" (Seiple 2007).

Finding a place to consider religion as a factor in Australian foreign policy and defence planning is critical.

5 Discussion

Given the significant influence of religion in Australia's near region, and Australian military and foreign policy assuming it is inconsequential, two directions are proposed.

5.1 Acknowledging Religion

In the first instance, acknowledging that religion is an important aspect in meeting Australia's interests in the region, efficiently addressing the challenges of contemporary crises and conflict, or even preventing initial outbreaks, means educating our strategic thinkers and senior military commanders. An awareness raising system that creates a common understanding through its inclusion of religion as a legitimate component of strategy across the key agencies of the Australian Government, allows Australian policy to provide a resilient and secure region. Building resilience, preventing conflict and, when required, winning conflict will take more than guns and will, by necessity, require a holistic approach that includes religion, that "missing dimension of statecraft" (Johnston and Sampson 1994).

Since there is no Defence agency currently filling the need, one way of addressing the significance of religion is to promote an Honorary Fellowship in Religion, Peace and Security at the Australian Defence Force Academy's main campus in Canberra—in line with the Defence Minister's recent appointment of an Honorary Fellowship in Women, Peace and Security (Department of Defence 2017). Additionally, suitably qualified ADF personnel might contribute to the enhancement of this important line of effort. A dilemma exists as to where this operational, rather than intelligence, advice might be drawn from—be that a generalist officer or an ADF Chaplain. The requirement to advise on targeting, shaping and influencing effects is essential. Suitably selected chaplains could be used because they are a resource in being, already tasked with providing advice to command as a specialist staff officer, consistent with their non-combatant status. According to doctrine, chaplains may conduct religious terrain assessments, enabling qualified advice to the commander and staff on the religious dynamics at play within the operational environment (CHAPMAN 2014).

5.2 *Training*

Secondly, understanding religion is central to a comprehensive effort to educate and synergistically apply all the elements of national power. In other words, the government needs members from across agencies that have been educated and trained together, and their agencies funded to understand how they act as a whole against the threat of "heretic terrorists" (Ali 2015). In this context, there needs to be a place for religion to be a part of the solution since it is so clearly a part of the problem (Seiple 2007).

Having an educated appreciation of religion throughout the spectrum of military operations will result in increased situational awareness, an improved preparation of the battlespace, enhanced force protection, and more effective tactical engagement. It is also likely to reduce risks. In addition, it creates opportunities contributing to the defeat—or success—of negotiations with opponents, an improved understanding of local populations and the underlying causes of conflict, and fruitful collaboration with allies (Field 2010). This will enable the commander to coordinate and focus the application of both physical and psychological effects, with a critical appreciation of the cultural characteristics and influences of all significant actors performing within the area of operation. Such capabilities can only enhance and contribute to the success of operations through risk reduction and the exploitation of opportunities, including the potential to influence behaviours and perceptions (British Ministry of Defence 2013).

One way of acknowledging the importance of training is to incorporate religious and worldview training as a component of the Defence Cooperation Program (DCP) and Special Operations Command bilateral exchange activities, in training conducted at the Australian Defence Force Peace Operations Training Centre (ADFPOTC), including religion as a major theme at the Australian Civil Military Centre, and training selected ADF chaplains in the area of command advice—specifically in conducting religious assessments, key leadership engagement, and engagement with other faith groups so that they can value add to an operational commander's plan in the area of Information Dominance and Influence (ID&I).

This type of training is currently being undertaken through an ad hoc approach across the Australian Army. Training includes religious and cultural briefings that go beyond 'cultural awareness' (Field 2010). Rather, soldiers are invited to engage in an interactive approach to cultural training that firstly highlights the importance or knowledge required to achieve lasting and enduring change. Secondly, they are asked to identify their inherent bias before finally considering the points of divergence that other people groups may have in order to identify opportunities for barrier removal, behaviour change and ultimately threat reduction. Anecdotally, soldiers and contingent commanders requesting the training believe it gives them increased confidence and competence in operating in a culturally complex environment.

6 Conclusions

Over the past 20 years several factors have converged making the issue of religion critical to military strategy. Given the uncertainty that comes with an interconnected world, religion offers people a reassuring meta-narrative that cuts against the uncertainty of the secular relativist worldview and provides a framework that can be used for socio-political influence. Australia, and the West more generally, has seen a rise in secularism, along with determined efforts to increase the separation of religion and state. This separation has prevented Australia from seeing its near region in a realistic manner. Nation states within Australia's near region, not to mention those further afield that the ADF are currently deployed into, are heavily influenced by spiritual or religious worldviews. Their worldviews are different to our own, leading them to think about the world differently, resulting in different behaviour to our own. Therefore, it is argued that understanding the place of religion, especially within Australia's near region, is an important aspect in meeting the challenges of contemporary conflict. This means removing inherent bias against incorporating religion in any assessment, grasping opportunities when religious engagement can be used for good, pursuing religious tolerance, and greater education for practitioners on the relevance of religion in strategic planning and practice.

Disclaimer The views expressed are the author's and not necessarily those of the Department of Defence. The Commonwealth of Australia will not be legally responsible in contract, tort or otherwise, for any statement made in this paper.

References

Ali, A. H. (2015). *Heretic: Why Islam needs a reformation now* (p. 63). Sydney: Harper Collins Publishers.

Arif, H. (2013). Islam – The global perspective. Circle of Peace. http://www.circleofpeaceonline.org

Barno, D. (2006). Challenges in Fighting a Global Insurgency, *Parameters* Summer 2006. http://www.carlisle.army.mil/usawc/parameters/06summer/barno.htm

Barton, E. (1897). Address to the National Australasian Constitutional Convention Debate Session II, Adelaide. In C. E. Bristow (Ed.), *Official report of the National Australasian Convention Debates* (p. 1185). Adelaide, Australia: Government Printer.

Bedsole, T. K. (2006). Religion: The missing dimension in mission planning. *Special Conflict* 2006. pp 9–15. http://isme.tamu.edu/ISME07/Bedsole07.pdf. Accessed 11 Sept 2017.

Betzold, C. (2015) Adapting to climate change in small island developing states. *Climatic Change* December 2015 133(3):481–489.

British Ministry of Defence (2013). *Red Teaming Guide.* Second edition. https://www.gov.uk/government/uploads/system/uploads/attachment_data/file/142533/20130301_red_teaming_ed2.pdf. Accessed 11 Sept. 2017.

CHAPMAN. (2014). *Defence Chaplaincy Manual (CHAPMAN).* Canberra, Australia: Department of Defence.

Department of Defence (2017). Media Release 4 April 2017. Canberra, Australia. https://www.minister.defence.gov.au/minister/marise-payne/media-releases/minister-defence-announces-new-visiting-fellowship. Accessed 11 Sept. 2017.

Durward, R., & Marsden, L. (Eds.). (2016). *Religion, conflict and military intervention.* Abingdon: Routledge.

Evans, M. (2008). The closing of the Australian military mind: The ADF and operational art. *Security Challenges, 4*(2), 105–131.

Field, C. (2010) The Enterprising Soldier. *The Australian Army Journal* VII(3) Summer. https://www.army.gov.au/sites/g/files/net1846/f/aaj_2010_3.pdf. Accessed 11 Sept. 2017.

Fiji Bureau of Statistics. (2017). *Statistics: Social Statistics: Religion.* Suva, Fiji. http://www.statsfiji.gov.fj/statistics/social-statistics/religion. Accessed 11 Sept. 2017.

Johnston, D., & Sampson, C. (Eds.). (1994). *Religion: The missing dimension of statecraft.* Oxford: Oxford University Press.

Kilcullen, D. (2005). Countering global insurgency. *The Journal of Strategic Studies, 28*(4), 597–617.

Lunn, J. (2009). The role of religion, spirituality and faith in development: A critical theory approach. *Third World Quarterly, 30*(5), 937–951.

Mandryk, J. (2010). *Operation world: The definitive prayer guide to every nation.* Downers Grove: InterVarsity Press.

Noble, R. (2009). Beyond cultural awareness: Anthropology as an aid to the formulation and execution of military strategy in the twenty-first century. *The Australian Army Journal VI, 2.* https://www.army.gov.au/sites/g/files/net1846/f/aaj_2009_2.pdf. Accessed 11 Sept. 2017.

Noble, R. Volesky, G. (2017). Theater Land Operations: Relevant observations and lessons from the Combined Joint Land Force experience in Iraq. *Military Review: Online Exclusive* June 2017 http://www.armyupress.army.mil/Portals/7/online-publications/documents/Volesky-Theater-Land-Operations.pdf. Accessed 13 Sept 2017.

Nunn, P. (2017). Sidelining God: why secular climate projects in the Pacific Islands are failing. *The Conversation*. https://theconversation.com/sidelining-god-why-secular-climate-projects-in-the-pacific-islands-are-failing-77623. Accessed 11 Sept 2017.

Olthuis, J. (1989). On worldviews. In P. Marshall, S. Griffioen, & R. Mouw (Eds.), *Stained glass: Worldviews and social science* (pp. 26–40). Lanham/ Maryland: University Press of America.

Papua New Guinea Defence Organisation (PNGDO) (2013). *The Defence White Paper 2013: Defending PNG's prosperity*. The Independent State of Papua New Guinea, Port Moresby, PNG.

Pew Research Centre (2015). Size and projected growth of major religious groups in Asia-Pacific, 2010–2050. In: The Future of World Religions: Population Growth Projections, 2010–2050. Online at http://www.pewforum.org/2015/04/02/asia-pacific/143-2/ Accessed 11 Sept 2017.

Schaeffer, F. A. (1982). He is there and he is not silent. In *Complete Works of Francis A. Schaeffer, A Christian Worldview* (Vol. I, p. 279). Wheaton: Crossway Books.

Seiple, C. (2007). The role of Religion in Winning the Long War. Online at https://globalengage.org/news-media/from-the-president/from-the-president-the-role-of-religion-in-winning-the-long-war Accessed 11 Sept. 2017.

United States Army (2014). *Insurgencies and Countering Insurgencies*, FM 3–24 / MCWP 3–33.5, Section 3–15. Online at http://www.apd.army.mil/epubs/DR_pubs/DR_a/pdf/web/fm3_24.pdf. Accessed 11 Sept 2017.

Verma, V. (2002). *Malaysia, state and civil society in transition* (p. 101). London: Lynne Rienner.

Wunderle, WD. (2006) *Through the Lens of Cultural Awareness: A Primer for US Armed Forces Deploying to Arab and Middle Eastern Countries*. Combat Studies Institute Press, Fort Leavenworth, Kansas, USA. Online at http://usacac.army.mil/cac2/cgsc/carl/download/csipubs/wunderle.pdf.Accessed 11 Sept 2017.

Chapter 7
Geographies of Irregular Warfare

David Kilcullen

1 Before Conflict: Strategic Geography at the Turn of the Century

Until the late twentieth century, that is before the beginning of the cycle of conflict that began on 11 September 2001, Australian strategy was constrained by a geographical framework defined by the 1986 Dibb Review and the 1987 Defence of Australia (DOA) doctrine (see Department of Defence, 1987 and Dibb 1986). In this pre-conflict thinking about strategic geography, DOA viewed adversaries as state-based, and Australia's strategic environment was defined by the continent's physical geography. Threats would come "from or through the archipelago to the north," and whereas the intent and capability of regional states was too fluid to form a basis for planning, 'geography' was considered to be a constant, and would play the defining role. DOA defined Australia's area of direct military interest as its immediate region, with its northern approaches dominated by a Sea-Air Gap that air and naval forces would control to prevent hostile states entering Australian territory, while land forces hung back to defeat any penetration of the continent's maritime defensive layers. Threats mattered more as they came closer, in a purely spatial sense.

Obviously enough, DOA embodied a rather stunted understanding of "geography" as solely physical. It paid little attention to the geographical impact of changing military technology (which transforms the strategic relevance of terrain features by altering the weapons and mobility platforms available to access, attack and defend them). It viewed sea- and air-space through a telluric, continental lens (as a "gap") rather than as a maritime highway, energy and revenue source, manoeuvre space, and power-projection base. It took little account of Australia's global

D. Kilcullen (✉)
Senior Fellow, Future of War, New America Foundation, Washington, DC, USA

S. Pearson et al. (eds.), *Australian Contributions to Strategic and Military Geography*, Advances in Military Geosciences,
https://doi.org/10.1007/978-3-319-73408-8_7

economic and political interests, essential to the nation's survival as a trader, commodities exporter and energy importer. And it underestimated the impact of the region's changing human, political and economic geography. But by the end of the 1990s, the main problem with DOA was the increasingly stark mismatch between strategic theory and operational reality.

I have described this as "pre-conflict" thinking, because to a large extent the nation—and by extension the Australian Defence Force—considered itself to be at peace during this period. But of course, reality was sharply different for those of us who came of age in that period. From 1986 until 1999, though resourced only for a close-in layered defence of the sea-air gap, Australian joint expeditionary forces deployed to Fiji, Namibia, Iraq (twice), Cambodia, Somalia, Rwanda, Bougainville (also twice) and East Timor, peacekeepers deployed to Cyprus, Western Sahara, the Sinai desert, the Balkans, Lebanon, Israel, and the Iran/Iraq border, and Australian demining teams served in Afghanistan and Pakistan. Only a few of these operations occurred in the "area of direct military interest"—they reflected Australia's global strategic interests, rather than its narrowly territorial defence requirements—and virtually all dealt with non-state actors and weak or collapsed (rather than functioning but hostile) states. In 1999 the crisis in East Timor offered a wake-up call, but it was the jihadist attacks of September 2001 that transformed Australians' understanding of strategic geography.

2 The Strategic Pivot of 9/11

On 9/11, a small non-state group, from a remote and barely governed base in Afghanistan, but exploiting globalised communications technology and worldwide transport, finance and information systems, mounted the most spectacular terrorist attack in history. This attack prompted a pivot in western strategic thinking, leading the United States and its allies to deploy combat forces to Afghanistan then Iraq, disrupting the entire geopolitical and strategic framework of the Middle East, North Africa and eventually, through the impact of forced population displacement, of Europe. Arguably—with the possible exception of the assassination of Archduke Franz Ferdinand in June 1914, which triggered the First World War and thereby transformed the geostrategic environment in vast regions of Europe, Africa, Asia and the Middle East—9/11 was thus the most consequential act of terrorism in recent history. It triggered a similarly global conflict and, indirectly through the Arab Spring, the wars in Syria and Libya, the rise of independent Kurdistan and the European immigration crisis, generated similarly transformative geographic realignments.

The 9/11 attacks also turned Australian thinking on strategic geography—as enshrined in DOA—on its head. For a detailed survey of this strategic history, see Evans (2001). Distant events turned out to have immediate, dangerous consequences through their power to radicalise local populations, transform national borders and disrupt global economic and energy systems, while the traditionally geographical distinction between domestic and international space, territorial zones of peace and war, or policing and military action, became increasingly ephemeral. The twentieth century strategic notions that threats would be primarily state-based, and that proximity equalled significance, turned out to be dangerous oversimplifications at the strategic level.

3 Strategic Geography in Wartime: The Operational Environment after 9/11

Within the context of this strategic realignment, at the operational level (the level of deployed forces conducting campaigns), the environment after 9/11 was marked by a complex geographical pattern of urban, littoral, networked operations against irregular (i.e. non-state) adversaries. Iraq, in particular, saw a heavy urban concentration of violence, with most combat action occurring in and around (that is, within the influence footprint of) cities. Urban populations—technically educated, with access to electrical power and manufacturing facilities, and able to communicate through increasingly dense mobile telephone networks—proved a formidable recruiting and support base for insurgent and terrorist groups, with a far higher latent military capacity than rural peasants or tribespeople.

As Australian forces transitioned from a conception of defence shaped by the DOA doctrine, towards a more expeditionary model after 9/11, this shift from a primarily rural (and tropical or sub-tropical) operating environment toward increasingly urban (and, in the case of Afghanistan, desert or mountainous) strategic terrain, exposed weaknesses in preparedness. Australians were forced to cross, and reckon with, the urban-rural divide.

The urban-rural divide also proved increasingly influential as a predictor for both population behaviour and enemy action in places like Afghanistan, Iraq, Libya, Syria, Egypt and, eventually, Europe and the United States. Traditional notions of rural guerrilla warfare gave way to a complex mix of urban and peri-urban combat in places like Baghdad and its surrounding "belts", the transitional zone of commuter towns and agricultural settlement ringing the city. Guerrillas still sought cover, as they always had, but with sophisticated air- and space-based surveillance systems making it easier to spot fighters in rural environments, that cover was increasingly to be found in the dense urban terrain and complex human geography of the cities.

As a consequence, war in this century has been both mainly irregular and overwhelmingly urban. The urbanisation of conflict revealed the city as a system, exhibiting a distinct urban metabolism, with operational geographies defined by flow and function in time and space. Irregular (insurgent and terrorist) groups evolved a distributed style of warfare that employed a dispersed pattern of violent actions—from rural ambushes to urban bombings to attacks on peri-urban infrastructure and disruption of flows of people, money and goods among cities—with the goal of influencing the full breadth and depth of a rural, peri-urban and urban nodal matrix.

Not that combat in rural or remote environments (such as, say, the jungles of central Africa, the Syrian desert or the mountains of Afghanistan) is disappearing. Such combat is continuing, and is something professional militaries will need to take into account for the foreseeable future. But as a proportion of the whole, and most importantly as a decisive element in warfare, urban operations are increasingly significant.

The littoralisation of warfare, in the same timeframe, mirrored an increasing coastal tendency in human settlement patterns, with more than 80% of people on the planet living within 80 kilometres of a coastline by the year 2000. It also reflected the underlying reality that war, as an endemic—almost a defining—human activity, tends to happen primarily where the people are. Just as most armed engagements worldwide this century have occurred in urban settings, most have also taken place within striking distance of the sea.

The increasingly littoral nature of warfare revealed the emerging military-geographical construct of "littoral operating environments" as both technologically driven and almost infinitely elastic. Traditional notions of littoral areas as spatially-defined—comprising coastal waters, coastlines, coastal strips and their immediate hinterlands—were replaced, in military strategy, with a functional definition. In this conception, littoral zones represented those areas of sea-space that could be engaged with land-based weapon systems, those areas of land that could be engaged from the sea, and the aerospace and cyberspace surrounding them. By this definition, as longer-range weapons and longer-endurance transport systems appeared, a military littoral operating environment could expand in turn until it bore little if any resemblance to the original geographical concept.

This was illustrated at the outset of the war in Afghanistan, when in October 2001 a force of U.S. Marines seized an airfield lying hundreds of miles inland, arriving in assault helicopters from ships that were themselves hundreds of miles offshore in the northern Arabian Sea. In military terms—as an area of land targetable from the sea—a land-locked desert airfield (in a country that was itself land-locked) turned out to be fully littoral. Projection of military power onto land from a sea base, and vice versa, was soon reflected in Australia's 2001 operating concept "Manoeuvre Operations in a Littoral Environment" (MOLE) (JFADT 2004). It was not restricted to states: Pakistani terrorist group Lashkar e-Tayyiba mounted a raid in 2008 against littoral targets in Mumbai, the Tamil Tigers fielded a formidable coastal flotilla of fast attack boats, and the war in Iraq involved considerable riverine and coastal warfare. The MOLE concept was one of Australia's major contributions to thinking on strategic geography in this early period of the conflict: the document was quoted, copied, critiqued and adapted by most of Australia's major allies, and no doubt analysed in detail by potential adversaries.

4 The Virtual Micro-Terrain of Insurgent Netwar

If the emergence of urban and littoral operations defined twenty-first century warfare at the strategic and operational level, at the tactical level the emergence of networked, self-synchronised multi-dimensional swarms defined the new tactics. A style of warfare christened "netwar" by the American theorists John Arquilla and David Ronfeldt (2000) proliferated. Writing before 9/11, they had predicted the emergence of a form of distributed, networked combat by non-state actors; by 2010, with the explosion of communications systems (smartphones, the internet, fibre-optic networks and satellite systems) netwar had become a reality.

The new tactical approach decoupled tactical action from strategic geography, as did the responses to it—responses that increasingly included drones, cyber operations and special operations. Guerrilla groups could reach out to inspire terrorist attacks on the other side of the planet as a way or relieving pressure on their local operations. They could field small, multi-role, semi-autonomous teams to flood an objective with multiple swarming combat groups. Snipers, using remote-controlled (tele-operated) sniper systems, could set up shop in an internet café in Belgium while collecting information and killing opponents through a network of a dozen weapons on the streets of Syria and Iraq (Bunker and Keshavarz 2016). Rebel mortar teams could use Google Earth, smartphones and iPad-based compass and inclinometer apps to generate levels of precision indirect firepower previously confined to nation-states.

In countering netwar (as defined by Arquilla and Ronfeldt 2000), countries like the United States and its allies increasingly developed tools for remote stand-off engagement, further decoupling combat action from geographical proximity. Ground crews in Africa could launch a drone that would be piloted by a remote station outside Las Vegas, Nevada, while receiving targeting guidance from a bunker in Hawaii, and firing approval from commanders in London, Kabul or Doha watching its full-motion-video feed, across multispectral sensors, in real time. Special operations teams in Italy and Spain could monitor and respond to developments in the Sahara Desert or along Libya's coastal strip. Officials in the situation room in the White House could watch in real time as special operators raided a terrorist compound on the other side of the planet.

At the level of insurgent micro-terrain, the emergence of human and electronic networks of action as key military objective gave rise to virtual theatres of war, based on human and informational geography rather than physical proximity, and so brought the post-9/11 transformation of strategic geography to its current state.

5 Conclusion: Exhaustion and its Implications for the Future

By 2017, after 16 years of continuous warfare, the political and national appetite for continued large-scale combat operations overseas was all but exhausted, in Australia as in the United States or other allied countries. What the French theorist Michel Foucault called "boomerang effects"—the geographical blowback of expeditionary operations and overseas occupations into home society in the form of militarised police, pervasive surveillance, mass migration and increased terrorism threat—was in full swing. The human cost of protracted overseas deployments—not to mention the economic impact of the 2008 global financial crisis—had made protracted large-unit expeditionary operations politically unsustainable, ironically presaging a potential return to a Defence-of-Australia-like strategy.

The imperative, now and into the future, for Australian strategists and others, is to make the "light footprint" work. Tools such as drones and special operations are clearly one part of the solution. Increasing reliance on contractors, local partnerships and "virtual presence" using cyber operations and aircraft is another. But ultimately, the critical challenge for strategists is in figuring out how to secure terrain (human, physical and informational) without occupation. How, without large-scale or long-term physical presence in a given geographical area, can we further our strategic objectives and further national interests?

Now that the decoupling of combat action from geographical proximity is fully realised, a key question for strategic thinkers is how to flip the script: how can we turn the increasingly urban, connected, coastal and cluttered terrain of modern war into an advantage rather than a disadvantage? How can we get out of our post-9/11 defensive crouch to exploit the emergence of a new, human-constructed operating domain—the cyber-domain—as a spatial adjunct to traditional manoeuvre? Future strategic success may indeed lie in finding the answers to these questions, making them an essential focus of future research.

References

Arquilla, J., & Ronfeldt, D. (2000). *Networks and netwars: The future of terror, crime and militancy*. Santa Monica: RAND Corporation.

Bunker, R. J., & Keshavarz, A. (2016). *Terrorist and Insurgent Tele-operated Sniper Rifles and Machine Guns*. Foreign military studies office (FMSO), Fort Leavenworth, Kansas, USA. pp 1–40.

Department of Defence. (1987). *The defence of Australia*. Canberra: Australian Government Publishing Service.

Dibb, P. (1986). *Review of Australia's defence capabilities*. Canberra: Australian Government Publishing Service.

Evans, M. (2001). *From Deakin to Dibb: The army and the making of Australian strategy in the 20th century*. Land Warfare Studies Centre, working paper no. 113, Australian Army, Canberra, Australia. Online at https://www.army.gov.au/sites/g/files/net1846/f/wp113-from_deakin_to_dibb_michael_evans.pdf

JFADT. (2004). *Australia's Maritime Strategy*. Joint Committee on Foreign Affairs, Defence and Trade. Parliament of Australia, Canberra, Australia. Ch 2.

Chapter 8
Strategic Military Geographies in the South China Sea

Greg Austin

1 Introduction

We have been using the term geopolitics for so long to mean simply great power competition that its actual meaning associated with geography has faded from view. There are many views on what the term means, but at its core it is the linkage made between international politics and military power on the one hand, and geographical spaces on the other. The relative absence of politics in military geography at the tactical level of war (in actual combat operations) should not blind us to the centrality of politics at the strategic level of military planning about geography. In peacetime, geopolitics is the strategic conduct of states when positioning themselves around certain geographies for the possible advantages they confer in the event of war or for the projection of power.

2 Military Geography as National Security Intelligence

At the strategic level, military geography can best be understood as national security intelligence about geography: how does a specific geography affect the national security interests, including war planning, for a country or group of countries. Thus,

G. Austin (✉)
School of Engineering and Information Technology, UNSW Canberra at the Australian Defence Force Academy, Canberra, Australia
e-mail: g.austin@unsw.edu.au

S. Pearson et al. (eds.), *Australian Contributions to Strategic and Military Geography*, Advances in Military Geosciences,
https://doi.org/10.1007/978-3-319-73408-8_8

as much as we want military geography to be an objective physical reality, it can never be that. War is a social activity that adapts to geography, transforming its social meaning into military and political outcomes.

It is axiomatic therefore that the military geography of a physical region is dependent on many non-physical factors, including the military strategies, doctrines, force structure, capabilities, technology and diplomacy, of those countries which might be affected in national security terms by the specific region. At the strategic level, an approach to military geography that sees it as an objective physical reality is a reductionist and pseudo-scientific approach. Reliance on such an approach may help explain why Western governments have performed so badly in counter-insurgency wars in Asia.

But just as importantly, we can distinguish between the "military geographies" of a single set of physical features (such as the South China Sea) depending on the different types of military campaign in contemplation: major war, low intensity conflict, or peacetime posturing. Even within each of those sets, more differentiation is possible: one military geography for defence of disputed island territories against other claimants, and another military geography for a campaign of harassment of international shipping, even though both are forms of low intensity conflict. Thus, strategic military geography is not just "constructed" (both constructivist and subjective), it is also ephemeral, according to the historical context and the changing actions of states.

In this subjective vein, rejecting a permanent and objective characterization of a military geography at the strategic level, the most important element of the concept for a single state is whether the space involves protection of national territory from occupation or invasion by another country. By contrast, for countries whose territorial sovereignty is not affected by a specific geography but which have other strategic interests, such as the security of trade routes, the strategic military geography of a maritime space will be very different from the one imagined by those countries which directly border it and which have sovereignty claims there.

This inevitable divergence helps to explain the intensity of some controversies about the military relevance of certain maritime spaces (their strategic military geography). The competition is as much about definition and political propaganda as it is about objective realities. There is never going to be an objective military geography of the South China Sea that is shared by any two countries. This means that states will compete in their accounts of the strategic military geography of the South China Sea according to the dictates of larger international and national security objectives, as well as history and politics.

A country's perception of the boundaries of its territorial sovereignty and related spatial rights in the air and the sea are not only physically imaginable as lines on a map but are driven by its understanding of various influences, especially history, national identity, justice and law.

3 Contested Semi-enclosed Sea, Micro Islets

The strategic military geography of a vast ocean area in a semi-enclosed sea, such as the South China Sea, is therefore context-specific: bound by time and circumstance and unique to a single state's national security purposes and capabilities. To illustrate the diversity of possible variations in national security geography of the South China Sea, one might take the rarely-heard view that the greatest national security "reality" in the region is storage of 100 tonnes of nuclear waste on Taiwan's Orchid Island (Botel-Tabaco in Fig. 8.1). This is a volcanic island in one of the most fragile tectonic areas in the world, with large earthquakes quite common in the past decade and with possible tsunami activity of an unpredictable character. If the storage site were to be seriously disturbed, nuclear waste damage to the regional environment and human health could be substantial.

The purpose of mentioning this danger is to offer an unfamiliar risk as one possible illustration of many hypothetical international security threats in the region. Even though this is not a danger for which any country plans a national security response, arguably we all should.

In the same spirit, it may be worth recalling a little history. The Philippines' territory of the Batan Islands, as shown in Fig. 8.1, sits astride the Luzon Strait, and at the south of the Bashi Channel at the northeast edge of the South China Sea. These may be in military operational terms the most important of the small non-coastal islands in the entire South China Sea. The islands are quite large (210 square kilometres in total). In World War 2 (WW2), Japan used its forces in Taiwan to attack and occupy the US territory of Batan Island (in the American colony of the Philippines) at the same time as it attacked Pearl Harbour (though the dates were 8 and 7 December respectively because of time zone difference). The Philippines was then a colony of the United States, as Taiwan was for Japan.

Some 76 years later, there is no agreed international territorial boundary between the Philippines and China (Taiwan) in the Luzon Channel. The last boundary in force was agreed between Japan and Spain when they had colonial control of Taiwan and the Philippines respectively. But this undefined sea border is not actively disputed. In May 2016, the Armed Forces of the Philippines felt it was important to hold a flag raising ceremony on the northernmost island of the Batan group, Mavulis (known also as Yami island), to reaffirm its control as part of nationwide maritime sovereignty awareness activities (Manila Livewire 2016). The absence of an agreed maritime boundary in this area after 1947 (when the Philippines became independent) may be one explanation why the Republic of China government on the mainland published its first 11-dashed line which enclosed the islands of the South China Sea and extended through the Bashi Channel into the Pacific Ocean.

The remainder of the 11-dashed line on the 1947 Chinese map had other drivers. During WW2 (that ended in 1945), China (then controlled by the Nationalist Party) had been seeking firm Allied agreement that after the defeat of Japan the Allies would recognise Chinese sovereignty over the Paracel and Spratly Islands, though getting Taiwan back from Japan was arguably the bigger prize. The Communist

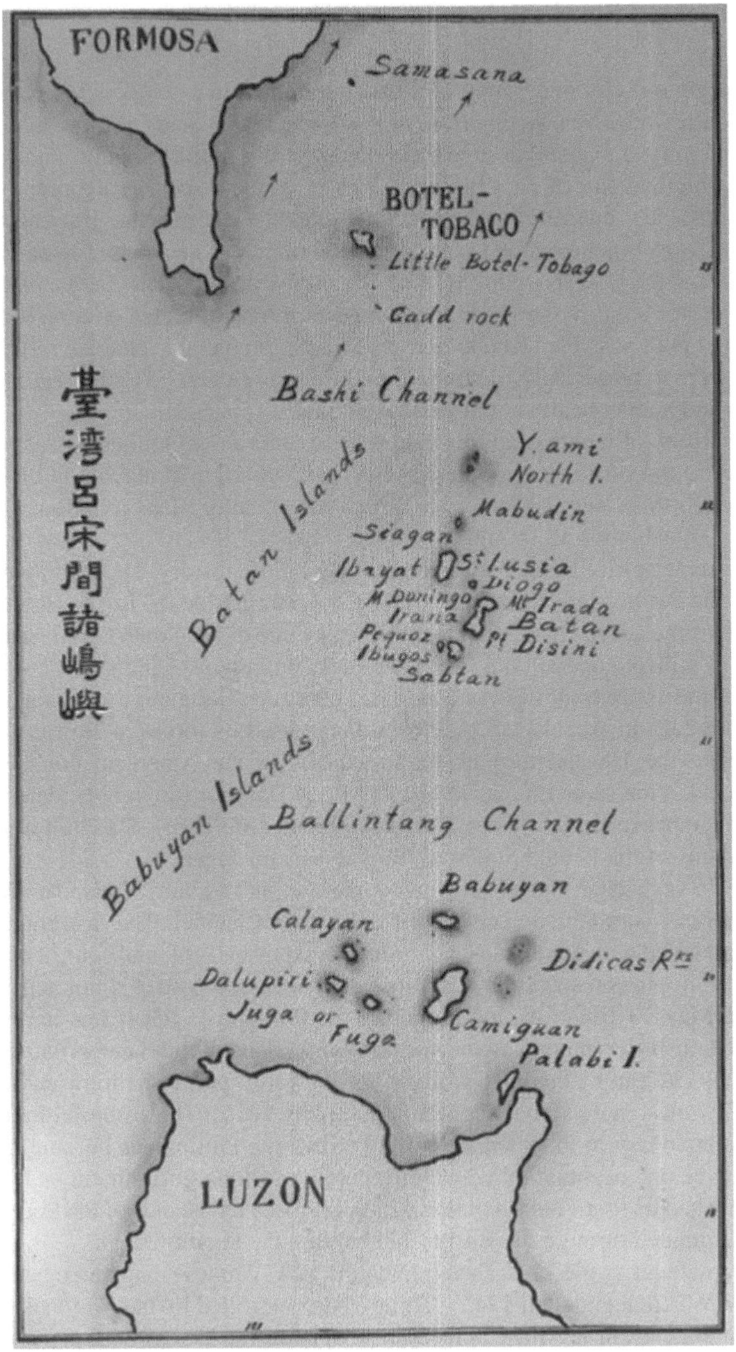

Fig. 8.1 Northern exit from the South China Sea. Source: Torii (1870–1953)

Party forces did not regain control of China's other big island of Hainan until 1950. The island of Taiwan is almost the same size as the Japanese island of Kyushu and almost three times bigger than the Philippines' island of Palawan; and about one third the size of the Philippines' island of Luzon.

Nationalist China could not succeed in getting a decisive verdict from the Allies, with the United States taking the view that it would be unwise to recognise China's sovereignty for geopolitical reasons regardless of the possible legality of its claims. Of some note, Japan—which had annexed the islands in 1939, 2 weeks before invading China's Hainan Island—did recognise Chinese sovereignty over the Paracel and Spratly islands in its peace treaty with China in 1951. The treaty was signed with the Republic of China (on Taiwan), which was at the time recognised by Japan as the government of all of China. Thus, the political geography of the South China Sea, including issues of boundary delimitation and island sovereignty, has been unsettled and subject to military activities since at least 1939 when Japan first annexed the Spratly Islands and invaded Hainan.

One cannot understate the visceral significance to China of this maritime region. What happens to the islands and reefs in the central and southern parts of the Sea is every bit as important as what happens in the Taiwan Strait and the Bashi Channel. China only regained sovereignty over one of its port cities on the South China Sea (Hong Kong) in 1997, barely 20 years ago. This is how China sees the political and strategic military geographies of the maritime space.

There are no Gibraltar-like islands in the mid-ocean of the South China Sea controlled by China or the United States. For comparison's sake, the land area of Gibraltar, a fortified territory that dominated access to the Mediterranean in WW2 is 6.8 km^2 (CIA 2017a) and it enjoyed land access through Spain for food and water during the war. And unlike the Mediterranean Sea, the South China Sea is in strategic terms, a porous maritime space, with no single country enjoying exclusive military control of strategic access points and all claimant countries able to host air force and missile capabilities reaching across the entire sea.

Compared with the islands of Hainan and Taiwan, the disputed islands in the central South China Sea (Paracel and Spratly Islands) are mid-ocean micro islets that do not have, even today after fortification of several of them, the same potential as Gibraltar had in WW2.

In terms of its share of China's maritime economy, the South China Sea occupies one third of China's highly-indented mainland coastline of around 18,000 km (Guangdong 4300 km and Guangxi Zhuang 1500 km). In addition, the island of Hainan presents a significant additional coastline in the South China Sea (1500 km). For Vietnam, the South China Sea represents the entirety of its maritime life. It seems odd to have to state it, but all littoral states (including China) should be accorded the status of equal participant in any strategic plans and activities in the South China Sea, and certainly ahead of the views of non-littoral and quite remote states. In fact, in contrast, China is regarded by many analysts as having no legitimate rights or interests in much of the South China Sea in spite of the country's status as a littoral country.

States of the region have been unable to agree all details of their maritime resource jurisdiction under provisions for continental shelf and exclusive economic zones in the 1982 Law of the Sea Convention, which only entered into force in 1994. Since 1947, before the Chinese Communist Party came to power, China has claimed an unusual martitime jurisdiction with a dashed line encompassing the South China Sea shown on national maps of the country. According to the US Department of State, China's current nine-dashed line encompasses about two million km^2 of the South China Sea area (Department of State 2014), which is commonly represented at 3.5 million km^2. This is vast maritime space, bigger in area than India. The southwest/northeast axis of the South China Sea (Singapore to Khaosiung) runs for about 3000 km—a little over half of the distance between London and New York—and about 50 to 90 hours sailing time for a warship, depending on the type. China's territorial claims to disputed islands initially encompassed a mere 13 km^2 of land area (prior to the equally small land reclamation on some reefs beginning in 2014).

The littoral countries understand only too well that the South China Sea is a semi-enclosed sea, and therefore a shared maritime space in the same way as the Gulf of Mexico, the Mediterranean Sea, the Baltic Sea, the Red Sea, the Persian Gulf, the Black Sea, the Sea of Japan, and the East China Sea. The Permanent Court of Arbitration in 2016 considered the South China Sea to be a semi-enclosed sea as defined by Part IX of the Law of the Sea Convention but failed to give it any weight in its deliberation. The full import of the two articles (122 and 123) in that chapter of the Convention remain unclear but they appear to convey the existence of shared responsibility for such an area without going as far as to assign "shared rights".[1] Littoral states of the South China Sea have not been able to come to a common understanding of what shared responsibility in a semi-enclosed sea means in practice because they cannot agree on the territorial claims.

The main military consequence for continental states that border a semi-enclosed sea (in this case China and Vietnam) is that their mainland coasts provide a substantial base for power projection into the sea area and they have land-based lines of communication to naval and air bases that can be used for naval operations and not just maritime lines of communication. In the case of the Gulf of Mexico, the United States and Mexico have overwhelming power projection potential in comparison with the much smaller island state of that semi-enclosed sea, Cuba. By contrast with Cuba in the Gulf of Mexico, the island state of the Philippines in the South China Sea provides a far greater landmass and potential power projection capability into that sea area than Cuba has, as does Malaysia, which has both a continental and island make-up. This means that most littoral states, except perhaps Brunei, have

[1] Article 123 says: "States bordering an enclosed or semi-enclosed sea should cooperate with each other in the exercise of their rights and in the performance of their duties under this Convention." That sentence seems to cover coordination of the exercise of all rights. The following sentence and sub-articles appears confined to coordination on resources exploitation, environmental management and scientific research, either "directly" (bilaterally) or "through an appropriate regional organization".

the potential to mount significant land-based attacks at sea and do not need to develop highly vulnerable air bases on tiny coral reefs to dominate the space.

4 Reshaping Physical Geography: From Reefs to Military Airfields

China has been developing artificial structures on submerged features in the Spratly Islands since 1988, when it moved to give a physical manifestation of its legal claims in place since at least 1945. There are several reasons why China delayed a physical move into its long-claimed territory (Austin 1998). One reason was that Taiwanese forces occupied one of the islands and for China, any military moves involving Taiwan are very sensitive. China and Taiwan were "at war" until 1979, when China made its first qualified peace overture toward the government there. Moreover, the United States and the Soviet Union had been very active in the region in the 1970s and 1980s, with a marked drop off in activity by the Soviet Union around 1986 and 1987. When it finally decided to move into the Spratly Islands, China was forced to occupy submerged features by building artificial structures because other claimants had occupied all of the natural islands in the Spratly group. After an initial armed clash with Vietnam in 1988, China over the next seven years seized submerged reefs to begin to establish a physical presence arguably to avoid clashes with the small military units of other claimants on the occupied natural islets.

The first Chinese structures built after 1988 in the Spratly Islands were comically inadequate. On Mischief Reef, in the mid-ocean, China built three connected platforms on stilts driven into the reef, each platform with a corrugated iron habitation built on it. The facility was manned by Chinese military personnel from the outset. Over time, China extended these artificial structures with pouring of concrete and making them gradually less flimsy and more habitable, including eventually the construction of a helicopter landing pad. In 1988, when China started to build its artificial structures, two rival claimants, the Philippines and Vietnam, already had rudimentary air strips on islets they occupied that could accommodate small fixed wing aircraft.

Later, Malaysia built an airstrip on one of the islands it claimed and had occupied prior to 1988, while the Republic of China (Taiwan) which had occupied Taiping Island first in 1946, and then again in 1956, built an air strip there in 2007.

All of China's most recent land reclamation (between 2014 and 2017) has been on submerged features in the vicinity of the Spratly Islands that have been the subject of Chinese physical presence for over two decades, so China has not permanently occupied new features for at least two decades. The suggestion by some that we have seen an "expansion" of Chinese claims is simply untrue. Chinese territorial claims have not changed since 1945, even though some analysts feared China would claim the artificial islands as generating a territorial sea and EEZ.

In September 2014, when new construction was revealed on Johnson Reef, a Chinese commentator echoed the Foreign Ministry by saying that the country needed to improve facilities there to make the lives of its occupants easier, but he added that China also needed an airbase to prepare for emergencies (Global Times 2014). He noted that this would be a breach by China of the 2002 Code of Conduct but that Vietnam and the Philippines had consistently breached it, and that China had shown great restraint by not using force to evict their forces from the islands and submerged features they occupied. On 9 April 2015, China acknowledged for the first time an unspecified military purpose to the reclamations which was almost certainly intended to mean a defensive position.

The Pentagon's considered assessment in 2017 of the military significance of the new moves by China on three of the features it has occupied for more than 20 years is worth noting. It says:

> "When complete, these outposts will include harbours, communications and surveillance systems, logistics facilities, and three airfields. Although artificial islands do not provide China with any additional territorial or maritime rights within the South China Sea, China will be able to use its reclaimed features as persistent civil-military bases to enhance its presence in the South China Sea significantly and enhance China's ability to control the features and nearby maritime space." (Department of Defense 2017)

It is hard to credit the extreme view published by CSIS that a 10,000-foot runway that China was building on Fiery Cross Reef "could enable China to monitor and potentially control the airspace over the South China Sea, which would provide greater capability to exert sea control" (Glasser 2015). This to me is hypothetically possible (in terms of monitoring) but less credible in terms of practical realities (controlling all South China Sea air space from one reef airfield against the might of the United States and its allies). This sort of overblown assessment reflects the mood in Washington and in US Navy circles that borders on rage at what they see (mistakenly) as China's destabilization. These sources seem unwilling to credit the possibility that actions by Vietnam, the Philippines and the United States itself might be contributing to the instability. In contrast, the US Department of Defence in 2017 used more temperate language, saying that Chinese base-building "enhanced the appearance of China's ability to exercise control over disputed areas in the South China Sea" (Department of Defense 2017). That is the appearance and not its actual capability.

On May 13 2015, US Assistant Secretary of Defense, David Shear, told the Senate Foreign Relations Committee that China's rival claimants, especially Vietnam, had been very active in extending their physical control (Shear 2015). Shear also pointed out that as of that time, China did not have an airfield as other claimants did. He said:

> "All of these same claimants have also engaged in construction activity of differing scope and degree. The types of outpost upgrades vary across claimants but broadly are comprised of land reclamation, building construction and extension, and defence emplacements. Between 2009 and 2014, Vietnam was the most active claimant in terms of both outpost upgrades and land reclamation, reclaiming approximately 60 acres. All territorial claimants,

with the exception of China and Brunei, have also already built airstrips of varying sizes and functionality on disputed features in the Spratlys." (ibid)

Shear (2015) was reporting that "Since 2014, China has reclaimed 2000 acres[2]—more land than all other claimants combined over the history of their claims". This comparison by Shear was at best irrelevant or at worst misleading. China was not "reclaiming" land around otherwise existing islands. It was from the outset extending artificial structures on submerged reefs. At the time of the Shear statement, the combined total land area of natural islands in the Spratly group—all occupied by other claimants—was already around 1200 acres (5 km^2), and that figure does not take into account any man-made extension of those land features through reclamation (CIA 2017b).

The "acres of concrete" debate, suggesting that each 100-acre increase in the total area changes the military balance, is somewhat fetishistic. It bears little relationship to bigger strategic realities in terms of national power confrontations. But the debate about acres is not irrelevant. It shows how deeply shocked all non-Chinese observers have been, even those more understanding of China's political position on the island claims, by the scale of the Chinese build of the artificial islands and their undoubted, if minimalist, military potential.

5 Control of the Geography: What China Sees When it Looks

China's military activities on its ocean frontier have given rise to a fear that it is seeking to expand its power at the expense of others now that it has a more powerful navy and much greater international economic power. The essence of this idea is that China's activities are expansionist and more aggressive compared with 20–30 years ago because it has a new urge for more territory or because it wants to throw its new-found weight around in maritime areas to rewrite regional order.

China sees it very differently. Its view is actually more in conformity with the facts, and less sinister. China's ocean frontier—its boundary with adjacent countries in the oceans—has for the most part never been settled in the five centuries since the idea of maritime borders under international law was first articulated in 1609. Between that time and the end of the imperialist and colonial era in 1945, China was either not in a position to delineate or not interested in delineating its maritime boundaries.

When the Communist Party gained power in 1949, it inherited the land territory and maritime territory of the Republic of China (ROC), although the legal and political status of the ROC and its territory (including maritime jurisdiction) is still not fully settled. Between 1949 and 1988, China fought a number wars over its land borders, falling into a pattern of negotiated settlement only after that time. China,

[2] 2000 acres is equivalent to 8.1 km^2, slightly larger than Gibraltar at 6.8 km^2.

with 22 countries bordering it on land or at sea, still has a significant land border dispute only with India. The settlement of its border disputes with former republics of the USSR, with which China nearly provoked a nuclear attack in 1969, occurred only in 1994, with final demarcation only occurring only in 2004. China's approach to maritime borders and maritime territorial disputes must be read against the background of this historical evolution. In border settlement terms, new China, not even 70 years old and gripped by massive internal disruption for the decade 1966–76, is still in the phase of demarcating its maritime borders.

One turning point that demonstrates this evolution is the decision by China in 1974 to attack and evict South Vietnamese forces in the Paracel Islands. Though China had maintained a claim to these islands during the latter period of France's colonial rule in Vietnam (1945–1954), it never moved to physically assert the claim as long as the Vietnam War was in full swing. Once the United States had withdrawn from the land combat in Southeast Asia in 1973, and with the South Vietnamese government faltering badly, China felt it faced little risk in attacking the small force on the Paracel islands.

China's primary motivation in occupying the Spratly Island features, including its significant military upgrading of them, is to defend what it sees as its island territories which it has formally claimed since at least 1945 (though it says since time immemorial). It sees the activities of other claimants as latter day attempts to usurp legitimate Chinese territorial rights. China's claims to land territory in the South China Sea did not change after 1949 when the Communist forces took power. China's claims to maritime rights in the South China have evolved since 1949 under the influence of evolution the Third UN Conference on the Law of the Sea (1973–1982) and under the influence of U.S. military pre-eminence on the China periphery.

As China sees it, regional order (the balance of economic and military power between Japan and China and between the mainland and Taiwan) has already been rewritten by China's peaceful rise (since 1990 at least) and any additional gains accruing from the control of its claimed small island territories in the South China Sea would be marginal. For China, the main game on its maritime frontier is successful unification with Taiwan which sits at the northern end of the South China Sea. Though China has come to describe the dispute in the Spratly Islands as a "core interest" because it involves sovereign territory, that is hardly new and is only a statement of the obvious. The more important characterisation driving Chinese policy for decades had been, as one Chinese government adviser observed in 1996, that the Spratly dispute is 'small in scale and local in nature'.

The extent and character of China's sovereignty claims are not unusual and, in broad terms, conform to the practice of other states with only one clear set of exceptions as noted: China appears to claim sovereignty over submerged reefs that would not normally qualify as land territory.

It is regularly asserted by some scholars, media commentators and other analysts that China claims sovereignty over almost the entire South China Sea. But that is based on a misunderstanding of the so-called nine-dashed line that China has repeatedly included in maps of the South China since 1947 and by China's ambiguity about the nature of the claims implied by the line. In December 2014, in a study of China's potential ocean frontier in the South China Sea, the US Department of State observed correctly that China had (up until that time) never clarified the jurisdictional intent of the U-shaped line (Department of State 2014). As discussed later, Chinese historic rights claims took on a much sharper edge after the 2016 decision of the Permanent Court of Arbitration. Yet, as the tribunal itself observed, the claims within the u-shaped line did not have a territorial character, otherwise China would not have established territorial sea baselines for Hainan and the Paracel Islands, which limit territorial sea jurisdiction to 12 nm from the baselines (PCA 2016: 91). The US DoD reported in May 2017 that the Chinese claims about the nine-dashed remain ambiguous (Department of Defense 2017).

China is still insisting it has a legitimate claim to historic rights of some kind in the nine-dashed line but this is indefensible, as the PCA tribunal found in 2016.

As argued in my 1998 book, *China's Ocean Frontier*, on which this chapter draws heavily, China is not helping itself by refusing to define its understanding of the line. But, as also noted in the book, the

> *"1947 publication of the unofficial boundary in the South China Sea was, in part, an effort by China to preserve its traditional or historic rights in the South China Sea in direct response to moves by other states, such as the United States in 1945, to extend their maritime jurisdiction for economic purposes beyond the previously accepted limits of the territorial sea"* (Austin 1998:208)

This can be understood as something of a Monroe doctrine intuition, as Scott (2016) has argued. Scott also notes that "An equally important motivation was the registering of China's claims to sovereignty over the islands within the u-shaped line" after the defeat of Japan and its retreat from the South China Sea.

Apart from the tumult of and territorial changes arising after the end of the Second World War, the South China Sea had been a place of territorial contest even earlier. Japan had been pushing forward in the South China Sea at the beginning of the twentieth century, France was the colonial power in Vietnam, the United States had become the colonial power in the Philippines in 1898, and the UK had even formally annexed two islands in the Spratly group in 1877. In 1902, Japan and the UK signed a security treaty directed against France and Russia. But, as noted by the Philippines in the PCA in 2016, China conceded in a diplomatic note to France in 1932 that at that time the Paracel Islands marked the southernmost extension of Chinese territory in the South China Sea (PCA 2016:82).

With the entry into force of the Law of the Sea Convention, clarification of historic rights by the International Court of Justice and by the PCA, and by the continuing practice of states, it will be almost impossible for China to claim (or prove) any historic rights that extend its maritime jurisdiction in the South China Sea beyond the twin regimes of exclusive economic zone (EEZ) or continental shelf as provided for in the Convention.

But China does have uncontested rights in the South China Sea, based on its recognized territory there (Hainan Island), its likely superior claim to the Paracel Islands and its equal claim to at least some of the Spratly Islands (see Austin 1998). States who want China to clarify the legal significance of its nine-dash line need to begin to understand and recognize that China does have some rights in the South China Sea and begin to work with China to define a mutually agreeable settlement.

But agreeing boundaries is not the root of the current military problem. China and the United States are more in dispute, along with other states, about what one particular type of maritime border means. The dispute concerns the right of non-littoral states to conduct military activities in the EEZ and the rights of non-littoral states to conduct (aggressive) patrolling for intelligence collection purposes in the EEZ. That is a bigger issue to which any final settlement of China's maritime frontier is now hostage and it clouds judgments outside and inside China of the lawfulness or otherwise of its positions.

Thus, we can conclude that the maritime territorial disputes predate the rise of China's power and increase in its naval capability. Any assumption that China has somehow expanded its maritime claims because it now feels more powerful is not borne out by the facts. One of many things that have changed about the disputes is China's willingness to act robustly, as most states would, to defend pre-existing sovereignty claims that have been in place for at least 72 years.

It is useful to compare China's massive use of force (an invasion) against Vietnam in 1979, an ally of the powerful USSR, with its relative restraint since 1988 in its maritime disputes with very weak neighbours, even as Chinese military power improved so greatly since 1998. In the war with Vietnam, which lasted from 1979 to 1988, China launched a large military campaign based on an initial assault to capture and temporarily occupy the five provincial capitals in the north. China withdrew after several months and after killing up to 100,000 Vietnamese, and then continued sporadic artillery bombardments for a decade. Chinese sources repeatedly claim that the war was forced on them by Vietnam's escalating border provocation and seizures of territory beginning as early as 1975, and surging in 1978. For an extended overview of China's motivations, see Zhang (2010).

It was in the 1970s that the Philippines made its first formal claims to any islands in the Spratly group even though it knew the islands had been claimed by China (and Taiwan) and Vietnam since at least 1945, and even earlier. Regardless of whether Vietnam or China had superior claims to the Spratly Islands that they had been contesting since at least 1945, the Philippines has unambiguously played the part of provocateur and latter-day mischief maker (Austin 2003).

6 Military Forces in the Semi-Enclosed Sea

One of the first questions to ask about China's view of the military geography of the South China Sea is its military intelligence picture of the region. China sees one great military contingency and two kinds of military threat. The great contingency is Taiwan (at the northern end of the sea). China no longer sees Taiwan as a military threat but the main mission of the Chinese armed forces is to prepare for the contingency of a military takeover of the island. Most western military analyses of the South China Sea ignore the biggest strategic issue. The relatively peaceful pattern of China's relations with Taiwan in the past 20 years does not fit the dominant Western narrative of an aggressive and militarising China, though China maintains a very powerful implied threat.

China sees a mid-level threat from the forward deployed U.S. forces in East Asia as a whole, including in defence of Taiwan and Japan, and therefore in potential combat operations in each of its littoral seas (the South China Sea, East China Sea and Yellow Sea). In contrast, China sees only a low-level threat arising from sovereignty and maritime disputes with some of its ASEAN neighbours, which have been characterised by very low levels of military activity for nearly four decades.

A missing link in many Western assessments of the geopolitics of the South China Sea has been consideration of the level of Chinese military activity in the area that might be considered normal given China's perception of its national security interests in the region. This assessment needs to be tempered by two additional and very different considerations: what is lawful and what might be considered a threat to peace. The discussion here does not address these three questions in detail. Instead it takes as a proxy for all three issues the notion of "military balance".

Since 2013, there has been a new round of alarmism in Asia about the strategic balance in the South China Sea. This concern stems in part from the mathematical comparison between Chinese and U.S. forces normally stationed in or near the South China Sea. China's order of battle in the area, in terms of numbers of ships, far outweighs that of the United States. This should be no surprise, given that China borders the sea area in question, and that it has legitimate and uncontested claims to something approaching 1.5 million square kilometres of maritime resource jurisdiction based on its mainland territory and Hainan Island. The fact that China also claims, and is militarily active around other small island groups, has to be seen in part at least (arguably the larger part) as subsidiary to the geopolitical reality that China is the most powerful state on the littoral of the South China Sea with lawful, uncontested jurisdiction (even from the Philippines' point of view) that extends to cover almost one third of the entire sea's area. (This calculation is based on a median line between Chinese territory recognised by the Philippines as such and Vietnamese and Philippines' territory, without giving EEZ effect to any of the Spratly Islands for any claimant, a consideration which the PCA confirmed as a proper one.)

The concept of strategic balance is a slippery one. It could be understood to mean anything between a comparison of orders of battle (a nearly useless analytical tool) and the much broader, highly political concept of balance of power. The idea of taking discrete theatres of potential conflict, such as the Taiwan Strait, and assessing the military balance (comparative orders of battle) in each theatre is not new to military analysis. It figured in the Napoleonic campaigns of two centuries ago. But it is a highly discredited tool among serious scholars looking at strategic rivalries in peacetime.

This is because static comparisons of orders of battle say nothing about the strategic intent or out-of-theatre capability that the parties might possess. Comparisons of numbers of in-theatre forces cannot convey anything about the military readiness of the units counted and a country's surge capacity for warfare in the theatre, including through potential or existing military alliances. Moreover, in the era of space-based assets and cyber potentials, a Napoleonic approach of counting the forces you can see would be to miss the dominant determinants of warfare in the modern age, such as stand-off weapons, space-based targeting and the U.S. military doctrine of prompt global strike.

The alleged asymmetry of motive is equally open to question. Since 1975, the Philippines and Vietnam have (for the most part) refused to back down from their claims even when confronted with Chinese pressure to do so. In fact, the biggest waverer in motive has been the United States. But even it has maintained a position of strength in the region for almost the entire post-war period—some 70 years, including after it was put on notice in 1987 to leave its military bases in the Philippines by 1991. The U.S. position on the South China Sea disputes could not have been more strident against China in the past several years and also remain within the bounds of peaceful if it tried.

In the light of these considerations, one might easily argue that in broad terms there is no asymmetry of potential military capability in the South China Sea in China's favour. What is concerning to many is that China could move quickly to eject the small military forces of Vietnam and the Philippines without a timely preventive action by any state. That it has not done so since 1988 should be taken as some indication of its likely intention not to do so absent a material change of circumstances. But to see China's possession of this or that small reef in the South China Sea as indicative of the strategic balance between China and "the rest" (ASEAN, United States, and Japan) is hardly defensible.

Anyone arguing that a strategic imbalance exists in the geopolitical contest probably needs to find new evidence to support the case because nothing in the public record to date, judged by serious analytical tests, reveals an asymmetry of strategic motive or strategic capability between the two opposing sides of the geopolitical positioning now occurring in the South China Sea: China versus "the rest". If anything, China is outgunned militarily and politically. The Philippines' action in the Court of Arbitration against China on maritime boundaries in the South China Sea has been a masterful diplomatic victory over China.

7 Misrepresenting Sea Lanes

Many of the semi-enclosed seas of the world have seen intense strategic and military rivalry at various times since 1945 but none of those maritime rivalries has been the cause of war or, with one or two exceptions, a sustained threat to shipping in that time. The exception is the "war on shipping" in the Persian Gulf in 1987 and 1988. A second exception might be attacks by communist Chinese naval forces on a small number of nationalist Chinse ships in the several years immediately after 1949. One reason why there have been so few anti-shipping campaigns since 1945 is that sustained attacks on a country's seaborne trade would be tantamount to a declaration of war. This was the dynamic in play in 1962 when the United States was prepared to risk war with the Soviet Union, possibly even nuclear war, by enforcing a naval quarantine on Soviet merchant ships carrying ballistic missiles to Cuba.

The heat being generated outside China about its putative threat to commercial shipping in the South China Sea because of activity in the Spratly Islands is simply not grounded in reality or any sense of the modern history of such campaigns. One of the best in-depth analyses of this military problem is a Top-Secret CIA-published inter-agency memorandum on Soviet intentions and capabilities for interdicting sea lines of communication in a war with NATO (CIA 1981). This report followed, and put an end to, several years of debate on the subject at high levels of the U.S. governments. The key judgements were that sea lines of communication attack was a secondary mission of the Soviet armed forces even in war and that the Soviet Navy would need to degrade NATO capability to defend its shipping before they could expect much success from such a campaign (CIA 1981:x). And that was for wartime. At the height of Soviet power in 1981, the United States did not ever imagine that the USSR would attack NATO commercial shipping in peacetime.

The term "sea lines of communication" (SLOC) is often used in debates about China as a substitute for "commercial shipping" though in normal parlance the two terms are not completely synonymous. While lines-of-communication is a normal military term of art, with a perfectly legitimate maritime equivalent, it does not refer to specific geographies but the totality of a military logistic and communications chain to a combat zone (tactically) or a theatre (at the operational level of war) wherever it might be and wherever it might move. The concept of "line of communication" never meant anything approaching a fixed route. The term SLOC is not dissimilar to the concept of "sea lanes", which refer not to defined lines across the sea on a map but to a set of maritime routes in their totality.

There are many exit/entry points into the South China Sea. The main potential choke points for international shipping—based on current routing—are the Malacca Strait, the Sunda Strait, the Bashi Channel (190 km wide), and the Taiwan Strait. It should be noted at the outset that shipping need not pass through any of these routes if one or more were blocked. Large oil tankers en route from the Persian Gulf to Japan, Korea and China already bypass the South China Sea entirely.

The Bashi Channel (between the Philippines and Taiwan) or the Taiwan Strait would probably be a better place to attack shipping entering or leaving the South China Sea than the open ocean areas around the Paracel and Spratly Islands. So, China does not need the Spratly Islands to threaten north-bound shipping in the South China Sea. It could do so easily (if it wanted to) without controlling this disputed island group. China's Southern Fleet is headquartered in Hainan which sits in a commanding position opposite the Philippines in the area that overlooks the northernmost egress from this semi-enclosed sea. China's mainland province of Guangdong has 4300 km of coastline that forms one side of this sea egress. The distance between this coastline and the Philippines coast is around 800 km and this area is in relatively easy reach of China's maritime military assets.

Most of the tiny islands and submerged reefs in the Spratly group are more than 800 km from Hainan Island. Chinese military leaders would have to be mad before they used these remote and tiny islands as the foundation for an anti-shipping campaign. Any country wanting to mount a sustained attack against shipping would probably use well protected, mainland-based air and naval assets, supported by a secure supply chain, before it would use combat aircraft operating from an air strip built on a remote submerged coral reef.

China almost certainly imports as much oil through the South China Sea as Japan, and possibly more. According to BP, China's oil imports in 2013 were 282 million tonnes, compared with 178 million tonnes for Japan. Just what share of China's imported oil goes through the South China Sea is unclear, but these two countries have a near identical interest in protecting shipping in the semi-enclosed sea. According to the US government in 2015, at least 60 per cent of Chinese oil imports in 2014 came from countries whose oil exports would have to use maritime Southeast Asia to reach China. Moreover, China's economy is highly dependent on sea trade, with manufacturing employment in the main coastal cities highly dependent on seaborne trade (between China and Japan, and China and the United States).

The historical precedents for a campaign against commercial shipping in open ocean areas are extremely small in number since 1900. There have been none since Germany lost hundreds of submarines in WW2 in its attempt to shut down seaborne trade and block naval access to Britain. In the month of May 1943 alone, Germany lost 46 submarines. In fact, a modern campaign against civil shipping is judged by naval experts to be most effective as the vessels leave port or approach the destination ports, or narrow straits, rather than in open ocean areas (such as around the Spratly Islands). One example of this was the war on shipping in the Persian Gulf during the Iran-Iraq War. The South China Sea is 3.5 million km^2 in area, around 14 times larger than the area of the Persian Gulf and unlike the Persian Gulf is not a long, narrow strip of water with only one exit point.

In the current era of satellite and sea-bed surveillance of Chinese warships and long-range stand-off weapons, any of its combatants (surface or submarine) would be picked off by the United States and its allies before they could do much damage to Allied shipping (though there would be losses).

Moreover, the number and tonnage of ships involved in seaborne trade is today many times higher than it was in during WW2. The increase in world seaborne trade between 1940 and now has been around 1000% (a factor of ten), with year on year increases expected for decades to come. Thus, the situation in East Asia today, if we were contemplating a possible China SLOC threat compared with the Battle for the Atlantic, is (in very crude terms) ten times more merchant ships facing ten times fewer submarines. Even if the submarines are much more capable today, so are anti-submarine surveillance assets.

The South China Sea (via the Malacca Strait) is a route of convenience (and lower cost) for shipping bound for Japan from the Indian Ocean. If there was a threat in the South China Sea, all shipping could simply avoid the Malacca Strait, divert south of Java, pass through the Sunda or Lombok Straits, enter the Java Sea and then the Philippines Sea to the east of the Philippines, never having entered the South China Sea. The added distance and time would be more costly, but the viability of this additional route in a conflict would undermine the value of any anti-shipping campaign focused around the Spratly Islands. This route is fairly consistently used by many larger oil tankers.

One reason why the "SLOC threat thesis" has emerged is that officers of the PLA Navy have paid increasing attention to the issue themselves. The mission of sea lanes protection is now mentioned far more often in official Chinese documents than 10 years ago. China has itself become far more vulnerable to interruptions of its seaborne trade. But another reason is that PLA officers are using this argument, just like their Western counterparts, as an additional lever to win more money for the defence budget.

At the same time, China's leaders know, and we can find expressions of this view, that for any country in today's world, protection of shipping is not something that one country can deliver by itself. It has to be a shared international responsibility. That is what China's official doctrine of 2013 says and I believe this to be the prevailing view in its government. The Office of Naval Intelligence in the United States has publicly stated a similar view (ONI 2015).

China's leadership actions in the Spratly Islands are influenced most by the belief that the specks of sand, rock or reef are Chinese territory. The leaders have no expectation or belief that control of the islands will enhance the country's military power projection capabilities in a way that might enable a confrontation over the safety of shipping with the world's greatest naval power and its many allies.

China is beginning to learn the hard way, and it is still learning, that its obduracy and military gambits in the Spratly Islands do not remotely serve its interests – even if its claim to the entire Spratly group were defensible. Diplomacy by the United States is working. It is not served by exaggerating or inventing military threats, such as that related to commercial shipping.

In a military crisis with Taiwan, China would almost certainly consider some sort of naval blockade against it, including an implied threat to commercial shipping. But if any military action was taken against commercial shipping, it would be in areas close to Taiwan, not in areas affected by a small Chinese military presence in the Spratly Islands or the Paracel Islands. Such a blockade would most likely be a naval quarantine exercise (blockade) not a war on shipping that saw China sinking Japanese, American, Singaporean and Australian commercial ships or cargoes. A war between China and Taiwan has become less likely in the past 20 years, not more likely.

8 Managing China's Geopolitical Shock

China's trade partners in the Asia Pacific who are suspicious of its military activities don't understand what it wants on its ocean frontier. Or, perhaps more correctly, none of them really agree with China's geopolitical lens. That is partly because it suits each of them to exploit propaganda value out of China's actions for their own narrow interests, often seated in domestic politics, rather than participating in a constructive and collective strategic approach to international resolution of the differences. It is also caused, in part, by the fact that few countries in Asia have strong, independent academic or intelligence expertise on China's maritime policy.

This is not helpful when China itself has a fragmented decision-making process for maritime frontier policy in an international environment that is reeling from China's willingness to deliver a strategic shock (island building) and where China is displaying a lack of interest in genuine negotiation on the maritime disputes.

This is a dangerous constellation of risk and ignorance. A circuit breaker is needed, as is more sustained and serious scholarly assessment of China's policy. The escalating tension will not be reversed until interested parties can defuse the strategic shock caused by China's massive upgrading of pre-existing artificial islands through 2015.

States hostile to China's island building, while dealing with the shock, need to understand that China is drawing a very big red line. In 1986, China's most "liberal" Communist Party General Secretary, Hu Yaobang, made an unprecedented leadership visit to the Paracel Islands. He declared that China would never surrender an inch of its claims. The Chinese military plan to establish a physical presence on submerged features in the Spratly Islands was undoubtedly made around this time. With its island building in the past 18 months, China has demonstrated that it will never resile from its territorial claim to the Spratly Islands and that it will compete at every level of international relations, including armed conflict if necessary, to defend the claims. If states opposed to China's claims do not acquiesce in China's territorial intent, then we are clearly on a pathway to a serious military clash of some kind over the sovereignty issue. Given what we know about these countries opposed to China, that acquiescence will never come absent a military "lesson" from China.

The only hope for enduring peace in the Spratly Islands is for China to move from a posture of strategic shock to one of strategic experimentation directed at keeping the peace. For more than 20 years, scholars from China and elsewhere have proposed several practicable and peaceable compromises to the South China Sea disputes. These proposals provide fertile ground for stabilisation. China has surprised the world with its policy innovations, ranging from "one country, two systems" to special economic zones and membership of the Communist Party for business people. It is time for this spirit of flexibility to inform China's geopolitical perspectives on the South China Sea with a view to a closing of the perception gap.

References

Austin, G. (1998). *China's ocean frontier: International law, military force and national development*. Canberra: Allen & Unwin.

Austin, G. (2003). Unwanted entanglement: The Philippines' Spratly policy as a case study in conflict enhancement? *Security Dialogue, 34*(1), 41–54.

CIA. (1981). Soviet intentions and capabilities for interdicting sea lines of communication in a war with NATO. Interagency Memorandum. November 1981. https://www.cia.gov/library/readingroom/docs/DOC_0000261312.pdf

CIA. (2017a). Gibraltar. The World Factbook. Central Intelligence Agency, Virginia, USA. Online at: https://www.cia.gov/library/publications/the-world-factbook/geos/gi.html (updated 16 October 2017) [Accessed 26 Oct 2017].
CIA. (2017b). Spratly Islands. In: East & South Asia, The World Factbook. Central Intelligence Agency, Virginia, USA. Online at: https://www.cia.gov/library/publications/the-world-factbook/geos/pg.html (updated 16 October 2017) [Accessed 26 Oct 2017].
Department of Defense. (2017). Annual report to congress: Military and security developments involving the People's Republic of China 2016. Department of Defense, Washington, USA. pp 7, 17 Online at: https://www.defense.gov/Portals/1/Documents/pubs/2016%20China%20Military%20Power%20Report.pdf [Accessed 26 Oct 2017].
Department of State. (2014). Limits in the Seas, No. 143: China: Maritime Claims in the South China Sea. United States Government, Washington, USA. pp 1–4 Online at: https://www.state.gov/documents/organization/234936.pdf [Accessed 26 Oct 2017].
Glasser, B. (2015). On the Defensive? China explains purposes of land reclamation in the South China Sea. Asia Maritime Transparency Initiative, The Center for Strategic and International Studies, Washington, USA. Blog post 20 April 2015, online at: https://amti.csis.org/on-the-defensive-china-explains-purposes-of-land-reclamation-in-the-south-china-sea/ [Accessed 26 Oct 2017].
Global Times. (2014). Truth behind Chigua Reef Construction. 11 September 2014. http://www.globaltimes.cn/content/880864.shtml [Accessed 26 Oct 2017].
Manila Livewire. (2016). AFP raised a Philippine flag on the Mavulis Island to assert sovereign territory. 2 May 2016. http://www.manilalivewire.com/2016/05/afp-raised-a-philippine-flag-on-the-mavulis-island-to-assert-sovereign-territory/ [Accessed 26 Oct 2017].
ONI. (2015). The PLA Navy: New missions and capabilities for the 21st century. United States Office of Naval Intelligence, Maryland, USA. p 11 Online at: http://www.oni.navy.mil/Portals/12/Intel%20agencies/China_Media/2015_PLA_NAVY_PUB_Print_Low_Res.pdf?ver=2015-12-02-081233-733 [Accessed 26 Oct 2017].
PCA. (2016). Award. Permanent Court of Arbitration. Case N° 2013–19. In the Matter of the South China Sea Arbitration. 12 July 2016. https://pcacases.com/web/sendAttach/2086 [Accessed 26 Oct 2017].
Scott, S. (2016). China's nine-dash line, international law, and the Monroe doctrine analogy. *China Information, 30*(3), 296–311.
Shear, D. (2015). Statement of David Shear, Assistant Secretary of Defense for Asian & Pacific Security Affairs before the Senate Committee on Foreign Relations May 13, 2015. United States Senate Committee on Foreign Relations. pp 3–4 Online at: https://www.foreign.senate.gov/imo/media/doc/051315_Shear_Testimony.pdf [Accessed 26 Oct 2017].
Torii, R. (1870–1953). Northern Exit from the South China Sea. A map from the inter-war period. https://en.wikipedia.org/wiki/Balintang_Channel#/media/File:Luzon_Strait_n7184.jpg. [Accessed 25 Oct 2017].
Zhang, X. (2010). Deng Xiaoping and China's decision to go to war with Vietnam. *Journal of Cold War Studies., 12*(3), 3–29.

Chapter 9
Australia's Most Southern Shores: The Strategic Geography of Antarctica and the Southern Ocean

AJ Press

But a *passive* acceptance that a stable past means a stable future will lead to a complacency that is not in Australia's strategic interests. The Australian government recently launched the *Australian Antarctic Strategy and 20 Year Action Plan* that sets out Australia's strategic interests in the Antarctic and Southern Ocean and provides a framework for Australia's future investment in the region.

Academic discourse on military strategic geography and the Antarctic is sparse. For Australia, any discussion of the strategic geography of the Antarctic must include the context and importance of Australia's Antarctic territorial claim; Antarctica's role as a place of peace and science; and the current and emerging geopolitics of the region. This chapter looks at the issues confronting Australia in the Antarctic and Southern Ocean – from climate change to potential geopolitical instability – and explores some of the responses that Australia could take to future challenges in the region.

1 Introduction

Australia has a long and proud history of involvement in Antarctic discovery, exploration, science and diplomacy. The first Australian to travel to Antarctica was Louis Bernacchi, a Belgium-born Tasmanian astronomer, who went south to Cape Adare with Carsten Borchgrevink's Southern Cross expedition in 1898. These were the first explorers to overwinter on the Antarctic continent.

A. Press (✉)
Antarctic Climate and Ecosystems Cooperative Research Centre & Institute for Marine and Antarctic Studies, University of Tasmania, Hobart, TAS, Australia
e-mail: tony.press@acecrc.org.au

S. Pearson et al. (eds.), *Australian Contributions to Strategic and Military Geography*, Advances in Military Geosciences,
https://doi.org/10.1007/978-3-319-73408-8_9

Douglas Mawson led the epic Australian Antarctic Expedition to east Antarctica from 1911 to 1914; and in 1929 he returned to the Antarctic and explored the coast of east Antarctica between 160° and 45° East (the expedition also visited Heard Island in November–December 1929). On this second expedition, Mawson claimed a large part of east Antarctica "[i]n the name of His Majesty King George the Fifth..." (Rothwell and Jackson 2011). In 1933, the United Kingdom transferred the Australian Antarctic Territory to Australia (*Australian Antarctic Territory Acceptance Act* 1933; Bush 1982). In 1936, Australia proclaimed the Australian Antarctic Territory, bringing into force the Australian Antarctic Territory Acceptance Act.

Not long after the end of World War II Sir Douglas Mawson was again influential in convincing the Australian government to focus on its Antarctic domain. In 1946 the government agreed to begin a program of research and exploration in the Antarctic, and in 1947 established the Australian National Research Expedition (ANARE; Bowden 1997; Haward and Griffiths 2011). In late 1947 an ANARE team sailed for Heard Island, landing there on 11 December 1947, claiming Heard Island and McDonald Islands for Australia on 26 December 1947. Another Australian expedition sailed for the Antarctic in December 1947; and in February 1948, an expedition set off to establish a station on Macquarie Island (Bowden 1997).

In 1948, the Australian government consolidated the administration of its Antarctic activities by creating an Antarctic Section within the Department of External Affairs, soon to become the Antarctic Division (now, the Australian Antarctic Division of the Department of the Environment and Energy).

Significantly, in 1954 Australia established the first research base, Mawson Station, on continental Antarctica. Situated in the western sector of the Australian Antarctic Territory, Mawson Station is the oldest, and the longest continuously occupied, research base in Antarctica.

After World War II, Cold War tensions spread to the Antarctic with suspicion over Soviet intentions in the South and heightening friction between the USSR and the United States. Various proposals were put forward to deal with the "Antarctic problem" (Christie 1951) - but ultimately the advent of the International Geophysical Year (1957–58) provided momentum and focus for the negotiation of the Antarctic Treaty (Berkman et al. 2011). Australia was one of the twelve original signatories to the Antarctic Treaty, and, as a claimant state (one of seven), played a significant role in its negotiation and construction (Haward and Griffiths 2011).

2 The Antarctic Treaty System

The cold war tensions of the 1950s spawned a unique set of international legal agreements that we now call the Antarctic Treaty System. The origins of the Antarctic Treaty System, its influence on Australian strategic geography, and its geopolitical evolution are discussed here.

2.1 Antarctic Treaty

The Antarctic Treaty plays an important role in Australia's strategic posture. The Treaty's first article states that "Antarctica shall be used for peaceful purposes only" and prohibits military activities, the establishment of military bases; and weapons' testing (*Antarctic Treaty* 1959). The Treaty also prohibits "nuclear explosions" and the disposal of nuclear wastes.

Most importantly for Australia, Article IV of the Treaty protects Australia's assertion of sovereignty over the Australian Antarctic Territory. It states, in part:

> No acts or activities taking place while the present Treaty is in force shall constitute a basis for asserting, supporting or denying a claim to territorial sovereignty in Antarctica or create any rights of sovereignty in Antarctica. No new claim, or enlargement of an existing claim to territorial sovereignty in Antarctica shall be asserted while the present Treaty is in force.

Initially in 1958 12 States signed the Antarctic Treaty and now there are 53 signatories. Of these 53, 29 are Consultative Parties - a status that gives them the right to participate in decision-making (vote) in Antarctic Treaty Consultative Meetings (Antarctic Treaty Secretariat 2017). Aside from the original 12 signatories to the Antarctic Treaty, Consultative Party status is achieved by Parties demonstrating a substantial level of scientific research in the Antarctic Treaty area.

Increasing numbers of Parties brings both strength – through increased global adherence to the fundamental tenets of the Antarctic Treaty system; and challenges – through changing expectations and norms in a diversifying geopolitical environment.

Since the Antarctic Treaty came into force in 1961, two more significant international agreements for the region have been negotiated and entered into force: The Convention on the Conservation of Antarctic Marine Living Resources (*Convention on the Conservation of Antarctic Marine Living Resources* 1980); and the Protocol on Environmental Protection to the Antarctic Treaty (*Protocol on Environmental Protection to the Antarctic Treaty* 1991). These international agreements form the basis of the Antarctic Treaty System and consequently, the framework of international governance of the region.

3 Convention on the Conservation of Antarctic Marine Living Resources

The 1980 Convention on the Conservation of Antarctic Marine Living Resources was negotiated because of rising concerns about the impact that krill fishing might have on whales. This conservation convention covers the marine environment of the Antarctic Treaty area as well as some regions to the north of 60°S including Australia's exclusive economic zone around Heard Island and McDonald Islands (CCAMLR Secretariat 2017). The Commission established by the Convention, among other things, manages the fisheries in the area, and sets annual catch limits for various species. Australia was a key negotiator of the Convention, and is an active participant in the Commission. Australian fisheries in the CCAMLR area are an important economic activity. In 2014/15 the total allowable catch for toothfish in the Heard Island fishery was 4410 tonnes. Australian companies also fish in other parts of the Convention area (Australian Government 2016a). The Antarctic and sub-Antarctic fisheries are the most significant direct economic activities by Australian companies in the Antarctic.

As the body responsible for the conservation of marine living resources in the Antarctic, the CAMLR Commission plays the key global role in ensuring the long-term protection of the Antarctic marine ecosystem, the ecological sustainability of Antarctic fisheries, and the measures required to eliminate potential conflicts over Antarctic marine living resources.

3.1 Protocol on Environmental Protection to the Antarctic Treaty

On 22 May 1989, the Prime Minister of Australia, Bob Hawke, sent shockwaves through the Antarctic Treaty System by announcing that Australia would not sign the minerals convention for Antarctica that had been under discussion for a decade and a half, and that had been formally agreed by Antarctic Treaty Parties the previous year (*Convention on the Regulation of Antarctic Mineral Resource Activities* 1988, the Minerals Convention).This dramatic move by Australia came after increasing domestic pressure by environmental groups to ban mining in Antarctica and create an Antarctic 'World Park' (Jackson and Boyce 2011).

Australia, France and other countries then proceeded to convince the Antarctic Treaty Consultative Parties to negotiate a 'comprehensive environmental management regime' for the Antarctic Treaty area. In 1991, the Parties met in Madrid and agreed to the Protocol on Environmental Protection to the Antarctic Treaty (Madrid Protocol; *Protocol on Environmental Protection to the Antarctic Treaty* 1991). It came into force in 1998. (The Minerals Convention never entered into force).

The Madrid Protocol proclaimed Antarctica a "…natural reserve, devoted to peace and science"; prohibits mining; and establishes a set of rules and a decision-making regime to assess and manage the environmental impacts of the activities of Antarctic Treaty Parties and their nationals in the Antarctic Treaty area.

3.2 International Convention on the Regulation on Whaling

The 1946 International Convention on the Regulation of Whaling (Whaling Convention, *International Convention on the Regulation of Whaling* 1946) pre-dates, and is not part of, the Antarctic Treaty System. While there is a currently a moratorium on commercial whaling in the Antarctic, Japan continues to kill minke whales in the Southern Ocean under a self-granted permit issued under Article VIII of the Whaling Convention (see Press 2016). The continued whaling by Japan has been a source of diplomatic conflict between Australia and Japan and the subject of a successful legal challenge by Australia in the International Court of Justice (ibid); and on-water protests by the non-government environmental organisation Sea Shepherd.

4 Australia's Antarctic Program

Since the establishment of the ANARE in 1947, Australia has been continually involved in Antarctic research and discovery. In the1980s, an Australian Government Antarctic Advisory Committee proposed a set of 'strategic interests' for Australia's Antarctic involvement.

Subsequent Australian Government decisions confirmed these strategic interests, although they were not publicly visible. In 1998 the Australian Government published a list of "Goals" for the Australian Antarctic program (Australian Government 1998), however, it is worth noting that these public goals differed from the "Strategic Interests" that Cabinet had agreed in 1989 (Press 2014). In 2014, the Australian Government commissioned a report on Australia's Antarctic strategic interests (Press 2014); and in 2016, the Prime Minister released the *Australian Antarctic Strategy and 20 Year Action Plan* (Australian Government 2016b). The Prime Minister's forward says:

> The Government is delivering a new era of Australian Antarctic endeavour. It is timely to reaffirm our national Antarctic interests, and to set out a plan to protect and promote them. [Author's emphasis]

The *Australian Antarctic Strategy and 20 Year Action Plan* (2016) presents Australia's "national Antarctic interests" as:

- maintain Antarctica's freedom from strategic and/or political confrontation.
- preserve our sovereignty over the Australian Antarctic Territory, including our sovereign rights over adjacent offshore areas.
- support a strong and effective Antarctic Treaty system.
- conduct world-class scientific research consistent with national priorities.
- protect the Antarctic environment, having regard to its special qualities and effects on our region.
- be informed about and able to influence developments in a region geographically proximate to Australia.
- foster economic opportunities arising from Antarctica and the Southern Ocean, consistent with our Antarctic Treaty system obligations, including the ban on mining and oil drilling.

While most of these 'national Antarctic interests' are essentially the same as those previously adopted by the Australian Government (Press 2014), there are two significant exceptions. The first, is explicitly aligning Antarctic scientific research with "national priorities" (fourth dot point above). Even though this has been implicit (modern) practice through the operation of published science plans for Australia's Antarctic program (Australian Government 2011), the newly articulated interests allow for clear prioritising of science in the national interest.

The second difference is the inclusion of the new national Antarctic interest "support a strong and effective Antarctic Treaty system". The inclusion of this specific 'interest' is a clear public indication that the Australian government sees that its national interests strategically align with support for the international governance regime of the Antarctic region. This is an important geopolitical statement by Australia, signifying deep commitment to the Antarctic Treaty System and its norms.

When Douglas Mawson set off on his Australian Antarctic Expedition in 1911, one of his aims was to make meteorological observations. He was sure that the Antarctic played an important role in weather on the Australian continent. Mawson's prescience has long been demonstrated, and recent analyses of ice cores are now revealing the important climate and weather teleconnections between Antarctica and Australia (ACE CRC 2015).

The science that Australia does in Antarctica is important for Australia and the world. The Australian Antarctic Science Strategic Plan (Australian Government 2011), which governs the priorities for Australian science in Antarctica, has the following themes:

1. Climate Processes and Change
2. Terrestrial and Nearshore Ecosystems: Environmental Change and Conservation
3. Southern Ocean Ecosystems: Environmental Change and Conservation
4. Frontier Science

The plan says, specifically, of the first three themes:

Themes 1, 2 and 3 address the priority science needs articulated by government policy and resource management agencies. The research in these themes will be designed explicitly to link investment in monitoring, observational and experimental science with the required process studies, synthesis and integrative modelling, and finally to the provision of scientific input to policy makers, conservation and resource managers. There will be increased emphasis placed on delivering project and program outputs and products to the end user community.

The plan prioritises environmental management, ecosystem science and fisheries, as well as climate change science, and provides guidance to researchers and decision makers about the allocation of funds and resources to Antarctic and Southern Ocean research. The Australian Antarctic Science Strategic Plan is currently under review (as foreshadowed in the Australian Antarctic Strategy and 20 Year Action Plan).

The climate change science that Australia does in Antarctica is significant on a global scale and it provides a strong platform for international collaboration under the Antarctic Treaty.

5 Australian Defence White Papers

Antarctica is a relative new-comer to Australia's defence white papers. The last three white papers (in 2013, 2015 and 2016) have consistently referenced the Antarctic and its Australian interests. The newest Defence White Paper (Australian Government 2016c) concludes that there is little direct threat to the peaceful geopolitical settings of the Antarctic region:

> *The Australian Antarctic Territory faces no credible risk of being challenged in such a way that requires a substantial military response for at least the next few decades. It is in our interest to work with like-minded countries to prevent any militarisation of Antarctica which could threaten Australia's sovereignty over the Australian Antarctic Territory and its sovereign rights over its offshore waters.*

The 2016 Defence White Paper goes on to state Australia's support for the Antarctic Treaty System:

> *Australia is a strong supporter of the Antarctic Treaty System, which expressly prohibits any mining in Antarctica. Australia also strongly supports the Convention on the Conservation of Antarctic Marine Living Resources, which regulates fishing activity in Antarctic waters.*

The 2016 White Paper also reflects the importance of "maritime protection operations" (i.e. fisheries enforcement and action against IUU fishers) in Australia's exclusive economic zones around Heard Island and McDonald Islands. Importantly, Defence also forecasts a role in "niche support" to the Australian Antarctic Division's logistics, especially the use of heavy lift aircraft in Australia's Antarctic program. Between November 2015 and February 2016, an aircraft of the Royal Australian Air force successfully transported heavy cargo to Antarctica to support the Australian Antarctic Program (Australian Government 2016d). A senior Air Force official, Air Commodore Iervasi, said at the time, the Royal Australian Air Force can "enhance... Australia's logistical and scientific capabilities in Antarctica". Since the 2015/2016 season, Royal Australian Air Force logistical support has become a regular feature of Australia's Antarctic program.

This new capability is a significant contribution to Australia's Antarctic efforts and provides a higher level of capability to engage with other nations in Antarctic science and logistics cooperation (Australian Government 2016e).

6 Discussion: A Southern Strategic Posture

While the Antarctic Treaty has successfully kept Antarctica free from military confrontation for the past decades, that does not diminish the strategic importance of the region for Australia. Nor does it diminish the need for resources to be committed to strategic thinking and future engagement.

Some Antarctic commentators portend a difficult and somewhat dystopian Antarctic future with crises and clashes over resources and sovereignty, and increased 'militarisation'. Dodds (2013) wrote of 5 "inconvenient truths" about the Antarctic. They were that Antarctica was "a deeply disputed space"; that the Antarctic Treaty System was struggling to manage "resource exploitation"; that countries were investigating its "mineral potential"; that Antarctica is "heavily commercialised"; and that there is "a growing anxiety about the role of China in Antarctica". While each of these subjects generate both media and academic interest, they are neither inconvenient, nor are they wholly accurate descriptions of the geopolitical nature of Antarctica and the involvement of countries there. A more sober analysis than Dodds is given, also in 2013, by the Australian Strategic Policy Institute (Bergin 2013).

In recent years there has been much written about Antarctic 'nationalism' (for example Dodds 2017; and for a counter view see Press 2012). Two major components of this discussion have been about the 'hidden agendas' of Antarctic states, and the rise of 'new Antarctic states' such as China (Brady 2010). In Australia's case the recent public profiling by the Government of Australia's national Antarctic interests underlines the fact that, as a claimant state in Antarctica, the expression of its national interests is neither contrary to the provisions of the Antarctic Treaty nor belligerent. A strong engagement in Antarctic affairs and a clear focus on Australia's position as an Antarctic claimant state is an important part of Australia's strategic geography in a changing world.

There has also been recent academic and media comment about potential threats to Australia's Antarctic interests from countries such as China and Russia (Brady 2010 and 2017, ABC 2015, ABC 2017). Much of this attention has focussed on other countries' stated desires to exploit Antarctica's resources, and these countries positioning themselves for a future where either the current indefinite ban on mining is overturned (Press 2015), or Antarctic Treaty fails. But the ban on mining contained in Article 7 of the Protocol on Environment Protection to the Antarctic Treaty is an indefinite ban. The Protocol does allow for a review conference to be called in 2048 – but any provision of the Protocol, including the ban on mining can be reviewed at any time by consensus of the Parties to the Antarctic Treaty. The simple mathematics of overturning the mining ban makes it unlikely that even if a review conference was called in or after 2048, the ban on mining would be overturned if it was opposed by a significant proportion of countries that currently support the mining ban (Press 2015). There is also speculation by some commentators that Antarctica is becoming 'militarised' through stealth, especially through the positioning of satellites, and the installation of satellite-signal ground stations in Antarctica (Bateman 2013; Brady 2017). Press and Bergin (2017) have responded to these claims:

> *All scientific research carried out in the Antarctic and Southern Ocean relies heavily on satellite telecommunications and the acquisition of remotely sensed data. The presence of Antarctic satellite receiving stations isn't of itself a breach of the Antarctic Treaty. If, however, they were used in Antarctica to engage in military activities, that would breach the treaty.*

> *What's required to stop militarisation of Antarctica is transparent and full reporting under Article VII—the requirement to disclose the use of military personnel and equipment. The Antarctic Treaty's right of inspection should be more regularly exercised.*

All of the scientific research carried out in the Antarctic and Southern Ocean relies heavily on satellite telecommunications and the acquisition of remotely sensed data. The outstanding advances in recent years of our understanding of the role that Antarctica and the Southern Ocean play in climate change, and its direct impacts such as sea-level rise, have required data from satellites. The same can be said of our increasing understanding of the health of Antarctic and Southern Ocean ecosystems – satellites are now integral to the way we study the planet, and the way we understand how it responds to human impacts. It is too simplistic to simply assert that the presence of Antarctic satellite receiving stations, which could have a 'dual purpose' for military use, is a blatant breach of the Antarctic Treaty. What is really required to stop the threat of "increased militarisation" of the Antarctic is for all Treaty Parties to strictly adhere to both Article 1 of the Treaty – "Antarctica shall be used for peaceful purposes only" – and also to transparent and full reporting and disclosure of the use of military personnel and equipment. Increased inspection under its Article VII should go hand in hand with strong diplomatic action should any Party behave contrary to the provisions of the Antarctic Treaty System. Australia's most recent Defence White Paper (Australian Government 2016c) looks primarily to the north, east and west of mainland Australia. While it does refer specifically to Australia's strategic interests in the Southern Ocean and Antarctica, a simple word count could lead to the conclusion that Defence has little interest in the South, and that security in this region is not a high priority. Such a conclusion is wrong. The region to the south of the Australian continent does not require combat-ready Defence engagement, because it is subject to an international treaty regime that preserves Antarctica free from military conflict.

But Australia has many "non-combat" responsibilities in the region and these make up a significant component of its national interests. The Australian search and rescue zone (the biggest in the world), extends to the Antarctic and covers much of the Australian Antarctic Territory (Australian Government 2017). In recent years Australia has been engaged in a number of high profile rescues in the region (Bergin and Grant 2014), and increased fishing, shipping and aircraft movements will likely lead to more calls on Australian assets and capabilities for search and rescue in the future. On-sea protests, such as those previously carried out by Sea Shepherd, also have a significant potential to lead to the engagement of Australia and its government and civilian assets in search and rescue in the Antarctic.

Australia has been engaged since the 1990s in combatting illegal, unreported and unregulated (IUU) fishing in the region (Darby 2013; Australian Associated Press 2004). Australia has used air force, naval and customs assets in these activities, and has worked with other countries such as France, New Zealand and South Africa to combat IIU fishing in the Antarctic. While IUU fishing has dramatically declined in the region over the past decade, IUU fishing and its associated environmental impacts still pose a future threat to the Antarctic ecosystem, Australian fishing interests, and the geopolitical settings of the region.

Forster, writing for the (Australian) Centre for Defence and Strategic Studies, considered that "… as nations seek solutions to the challenges of climate change, energy, water and food security… Australia's national security interests in the Antarctic region are likely to be increasingly challenged over the next 20 years" (Forster 2016, abstract)). In the face of these potential challenges, Foster went on say that "… Australia will have to invest more in Antarctic research, enforcement operations, logistics, diplomatic and bureaucratic support" (Forster 2016, p11). Foster recommended that Australia "…needs to be visibly committed to a strong presence in the Antarctic [;] … needs to strengthen important relationships with emerging influential states [and]… needs to reinforce its leadership role in science and governance…" (Forster 2016, p 12).

Izaak Gurney, a serving naval officer looking at the use of Australian defence assets in the Southern Ocean recommended that defence heavy lift aircraft be dedicated to support Australia's Antarctic logistics; that new investment be made in dedicated Southern Ocean patrol vessels; and, inter alia, increased logistics collaboration with New Zealand (Gurney 2016). The White Paper also reflects the importance of "maritime protection operations".

7 Conclusions

Traditional views on strategic military geography are mainly about geographic places, but in the Antarctic, the military can play a key role in supporting scientific research. The use of military capability in support for Antarctic science and logistics is a productive addition to the operations of many national Antarctic programs, and does not contravene the Antarctic Treaty. From a whole-of-government perspective, use of national assets in the Antarctic is significantly more efficient than duplication of equipment and personnel. In Australia's case the operation of the C-17A 'Globemaster' provides new capability for heavy lift operations and increases the reach of Australia's Antarctic program. Combined with other traditional and emerging Australian Antarctic capabilities, the use of military logistics also increases the ability of Australia to collaborate more extensively with other Antarctic Treaty nations in science and logistics.

The Antarctic Treaty and its related laws and international agreements have proved remarkably resilient to political shocks. They have survived the Cold War; the Falklands/Malvinas crisis between the UK and Argentina; the political turmoil of Australia, France and others walking away from the agreed Convention on the Regulation of Antarctic Mineral Resource Activities; the divisive debates in the United Nations, led by Malaysia (1980s–2000s) about the status of Antarctica; and the growth and diversification of nations involved in the Antarctic Treaty System. There have been no 'Antarctic' armed conflicts and the explicit emphasis on science in the Antarctic Treaty provides a genuine focus for nations to engage constructively in Antarctic affairs.

Complacency, though, is not a wise policy setting. Geopolitical tensions in other parts of the globe do have the ability to spill over into Antarctic affairs. The rise of 'new' Antarctic nations has the potential to alter long established diplomatic and political norms in the Antarctic. Depletion of fisheries in other parts of the world has the potential to spark a new outbreak of IUU fishing in the Southern Ocean, and to see fishing nations attempt to dilute and circumvent the precautionary catch limits for Antarctic fisheries. While it is unlikely to happen in the near future, some countries within and outside the Antarctic Treaty System may also attempt to have the indefinite ban on mineral exploitation reviewed or overturned (Press 2015).

Australia's commitment to a "strong and effective" Antarctic Treaty System is the most robust defence against future shocks and crises in the Antarctic. It is the inherent characteristics of the system itself that provide the foundation for its strength, and it is the common objectives of peace and science that provide the framework for effective international cooperation in the region. As pressures on the rest of the planet increase through environmental degradation, overexploitation, and climate change, the importance of understanding and protecting the Antarctic increases. Australia's commitment to this region, therefore, should not waiver: understanding the Antarctic is critical to understanding the future of Australia; protecting Antarctica's environment and ecosystems is an investment in the future of our planet; and ensuring a strong international commitment to the Antarctic Treaty System is an investment in regional and global peace and security.

Working closely with its global Antarctic partners, Australia should ensure that it supports its Antarctic aspirations with a strong, focussed and well-funded Antarctic program, the integration across the whole of Government of Antarctic policy and operations, and a robust engagement in Antarctic diplomacy.

References

The Australian Broadcasting Corporation, ABC. (2015). http://www.abc.net.au/news/2015-01-20/chinas-desire-for-antarctic-mining-despite-international-ban/6029414 [Accessed 20 July 2017].

The Australian Broadcasting Corporation, ABC. (2017). http://www.abc.net.au/news/2017-05-26/australia-urged-to-spend-more-on-antarctica-program/8561624 [Accessed 20 July 2017].

The Antarctic Climate and Ecosystems Cooperative Research Centre, ACE CRC. (2015). *Position Analysis: Ice cores and climate* at http://acecrc.org.au/wp-content/uploads/2015/08/2015-ACE-PA-Ice-Cores-and-Climate.pdf (Accessed 30 May 2017).

Antarctic Treaty. (1959). At http://www.ats.aq/documents/ats/treaty_original.pdf (Accessed 31 May 2017).

Antarctic Treaty Secretariat. (2017). http://www.ats.aq/devAS/ats_parties.aspx?lang=e (Accessed 30 May 2017).

Australian Antarctic Territory Acceptance Act. (1933). Australia, at https://www.legislation.gov.au/Details/C2004C00416 (Accessed 31 May 2017).

Australian Associated Press. (2004). Toothfish pirates to face armed patrols, *The Age* http://www.theage.com.au/news/Breaking-News/Toothfish-pirates-to-face-armed-patrols/2004/11/19/1100838216395.html (Accessed 5 June 2017).
Australian Government. (1998). *Our Antarctic Future: Australia's Antarctic Program Beyond* 2000. The Howard Government response to Australia's Antarctic Program Beyond 2000: A Framework for the future (A report to the Federal Government by the Antarctic Science Advisory Committee), Canberra.
Australian Government. (2011). *The Australian Antarctic Science Strategic Plan 2011–12 to 2020–21* at http://www.antarctica.gov.au/__data/assets/pdf_file/0019/27307/AASSP_final-published-version_Apr-2011.pdf (Accessed 5 June 2017).
Australian Government. (2016a). *Australian Fisheries Status Report 2016* at http://data.daff.gov.au/data/warehouse/9aam/fsrXXd9abm_/fsr16d9abm_20160930/00_FishStatus2016_1.1.0.pdf (Accessed 31 May 2017).
Australian Government. (2016b). *Australian Antarctic Strategy and 20 Year Action Plan* at http://www.antarctica.gov.au/__data/assets/pdf_file/0008/180827/20YearStrategy_final.pdf (Accessed 30 May 2017).
Australian Government. (2016c). *2016 Defence White Paper*, Department of Defence, Canberra at http://www.defence.gov.au/whitepaper/docs/2016-defence-white-paper.pdf (Accessed 30 May 2017).
Australian Government. (2016d). C-17A Antarctic flights prove a huge success. At http://www.antarctica.gov.au/news/2016/c-17a-antarctic-flights-prove-a-huge-success (Accessed 18 Sept 2017).
Australian Government. (2016e). *Deep-field air drop supports Antarctic science* at http://www.antarctica.gov.au/news/2016/deep-field-air-drop-supports-antarctic-science (Accessed 5 June 2017).
Australian Government. (2017). Australia's search and rescue obligation, https://www.amsa.gov.au/search-and-rescue/ (Accessed 5 June 2017).
Bateman, S. (2013). *Is Antarctica demilitarised?* At https://www.aspistrategist.org.au/is-antarctica-demilitarised/ (Accessed 24 July 2017).
Bergin, A. (Ed.). (2013). *Cold calculations: Australia's Antarctic challenges* Australian Strategic Policy Institute. https://www.aspi.org.au/publications/strategic-insights-66-cold-calculations-australias-antarctic-challenges (Accessed 5 June 2017).
Bergin, A., & Grant, D. (2014). *Search and rescue: a growing responsibility* Australian Strategic Policy Institute at https://www.aspistrategist.org.au/search-and-rescue-a-growing-responsibility/ (Accessed5 June 2017).
Berkman, P. A., Lang, M. A., Walton, D., & Young, O. R. (2011). *Science diplomacy: Antarctica, science and the governance of international spaces* (1st ed.). Washington: Smithsonian Institution Press.
Bowden, T. (1997). *The Silence Calling* (1st ed.pp. xx–xxvii). Sydney: Allen and Unwin.
Brady, A.-M. (2010). China's rise in Antarctica? *Asian Survey, 50*(4), 759–785.
Brady, A.-M. (2017). China's expanding Antarctic interests: Implications for Australia. Australian Strategic Policy Institute, Special Report. https://www.aspi.org.au/report/chinas-expanding-interests-antarctica (Accessed 18 Sept 2017).
Bush, W. M. (1982). *Antarctica and international law: a collection of inter-state and national documents* (Vol. 2, 1st ed. p. 143). New York: Oceana Publications.
Christie, E. W. H. (1951). *The Antarctic problem: an historical and political study*. Sydney: Allen & Unwin.
Commission for the Conservation of Antarctic Marine Living Resources. (2017). At https://www.ccamlr.org/en/system/files/CCAMLR-convention-area-map-large.pdf (Accessed 30 May 2017).
Convention on the Conservation of Antarctic Marine Living Resources (1980). At https://www.ccamlr.org/en/document/publications/convention-conservation-antarctic-marine-living-resources (Accessed 31 May 2017).

Convention on the Conservation of Antarctic Marine Living Resources Secretariat. (2017). At https://www.ccamlr.org/en/system/files/CCAMLR-convention-area-map-large.pdf (Accessed 30 May 2017).

Convention on the Regulation of Antarctic Mineral Resource Activities. (1988). At https://www.ats.aq/documents/recatt/Att311_e.pdf (Accessed 5 June 2017).

Darby, A. (2004). Toothfish pirates to face armed patrols. *The Age* November 19, 2004 http://www.theage.com.au/news/Breaking-News/Toothfish-piratesto-face-armed-patrols/2004/11/19/1100838216395.html (Accessed 5 June 2017).

Darby, A. (2013). Ocean pursuit leaves toothfish pirates with nowhere to run, *Sydney Morning Herald,* http://www.smh.com.au/environment/animals/ocean-pursuit-leaves-toothfish-pirates-with-nowhere-to-run-20130503-2iyl1.html (Accessed 5 June 2017).

Dodds, K. (2013). *Five inconvenient truths about the Antarctic* at https://blog.oup.com/2013/03/five-inconvenient-truths-about-the-antarctic/ (Accessed 5 June 2017).

Dodds, K. (2017). Awkward Antarctic nationalism: Bodies, ice cores and gateways in and beyond Australian Antarctic Territory/East Antarctica. *Polar Record, 53*(1), 16–30.

Forster, M. (2016). *Australia's national interests in the Antarctic region: What is important?.* Canberra: Centre for Defence and Strategic Studies. At http://www.defence.gov.au/ADC/Publications/IndoPac/Forster_IPSP_Final.pdf (Accessed 24 July 2017).

Gurney, I. (2016). Cold fleet: The Southern Ocean, Antarctica and the Australian defence force, Canberra: Centre for Defence and Strategic Studies. At http://www.defence.gov.au/ADC/Publications/Commanders/Gurney_paper.pdf (Accessed 24 July 2017).

Haward, M., & Griffiths, T. (Eds.). (2011). *Australia and the Antarctic treaty system: 50 years of influence* (pp. xx–xxvii). Sydney: UNSW Press.

Jackson, A., & Boyce, P. (2011). Mining and 'World Park Antarctica' 1982–1991. In M. Haward & T. Griffiths (Eds.), *Australia and the Antarctic Treaty System: 50 years of influence* (pp. 243–273). Sydney: University of New South Wales Press.

Press, A. J. (2012). Strategic bifocalism: National views in an international context. In J. Jabour, M. Haward, & A. J. Press (Eds.), *Proceedings of the symposium to mark 75 years of the Australian Antarctic Territory* (1st ed., pp. 115–119). Hobart: Institute for Marine and Antarctic Studies.

Press, A. J. (2014). *20 Year Australian Antarctic Strategic Plan* at http://www.antarctica.gov.au/__data/assets/pdf_file/0008/178595/20-Year-Plan_Press-Report.pdf (Accessed 30 May 2017).

Press, A. J. (2015). The Antarctic treaty system: Future mining faces many mathematical challenges. In *Year Book of Polar Law, 7,* 623–631.

Press, A. J. (2016). Science in the Court! The role of science in "Whaling in the Antarctic". In M. Fitzmaurice & D. Tamada (Eds.)., (2016) *Whaling in the Antarctic: Significance and Implications of the ICJ Judgment* (pp. 346–386). Amsterdam: Brill.

Press, A. J., & Bergin, A. (2017). China on ice. *The Strategist.* At https://www.aspistrategist.org.au/readers-response-china-ice/ (Accessed 18 Sept 2017).

Protocol on Environmental Protection to the Antarctic Treaty. (1991). At http://www.ats.aq/e/ep.htm (Accessed 7 June 2017).

Rothwell, D., & Jackson, A. (2011). *Sovereignty.* In M. Haward & T. Griffiths (Eds.), *Australia and the Antarctic Treaty System: 50 years of influence* (pp. 48–67). Sydney: University of New South Wales Press.

[illegible]

Chapter 10
The Evolution of Geospatial Intelligence

Robert S. Coorey

> Know the enemy, know yourself; your victory will never be endangered.
> Know the ground, know the weather; your victory will be total.
>
> Sun-Tzu (544–496BC)

1 Introduction

Knowledge and understanding of the enemy's location, intent, size and capability of its forces and battle ground topography have been fundamental to battlefield decision making since the earliest of times. Military geography is the term used to describe this area of geographic knowledge. Over time the application of military geography has changed as weaponry, economics, social and geopolitical influences have altered the scale and dynamics of military operations.

Today, the scope of military geography includes social interaction, natural and cultural features on the earth's surface, together with features above and below it.

Geospatial intelligence (GEOINT) forms the base of the Common Operating Picture (COP) on which all other information and intelligence is layered. To generate GEOINT, analysts collect, collate and synthesise geospatial data from a wide range of sources. The GEOINT, as part of the COP, forms the military geography landscape to enable the commander to consider the increasingly complex battlespace.

Today, geospatial data is constantly collected—in both a civilian and military context. Increasingly, the rate at which such data are collected forms the crux of the big data issue for the military geographer. One remote sensing mission can collect gigabytes of data and this quickly runs into terabytes and petabytes as the data is

R. S. Coorey (✉)
Geospatial Intelligence Pty Ltd, Kingston, ACT, Australia
e-mail: rcoorey@geoint.com.au

S. Pearson et al. (eds.), *Australian Contributions to Strategic and Military Geography*, Advances in Military Geosciences,
https://doi.org/10.1007/978-3-319-73408-8_10

archived for future reference. Likewise, with billions of social media users, the volume of potentially useful data generated each day is immense.

The art of GEOINT has changed over time from simply fusing and integrating data for a COP into connecting, sorting, filtering, prioritising and visualising data to provide useful, reliable and actionable intelligence. GEOINT will continue to change and adapt to the changing scale and dynamics of military operations.

This chapter discusses the past and present roles that GEOINT plays in strategic and military geography. In discussing the role of GEOINT, a fundamental characteristic will be highlighted, that of the evolution of the use and role of data and information in informing the strategic and operational elements of military geography. The future role of GEOINT encompasses the development and application of advanced technologies, which synthesise multi-dimensional geographic reference data and seamlessly present the results intelligibly on a variety of devices in real time.

GEOINT has become a science in its own right - combining the disciplines of imagery analysis, remote sensing, geographic information systems, spatial science, and land management as well as social media analysts.

2 Past

The study of geography and how it relates to military strategy is not new. Sun-Tzu (544-496 BC) dedicated the bulk of the *Art of War* to geography, tactics, and strategy (Sun-Tzu and Griffith 1964). He argued that a commander needs ‘full knowledge’ of the terrain to ensure victory. Sun-Tzu also espoused the wisdom that strategic territory was worth making alliances, treaties and agreements over.

Carl von Clausewitz in *On War* (Clausewitz et al. 1984) advised that a line of operations is more dependent on geography than geometry. In this case, geometry refers to the lines military planners would draw on maps without regard to the situation on the ground. Clausewitz was explaining that terrain shapes operations and the effects of weather on terrain.

Thomas Maguire was perhaps the first to write an entire book on strategy and military geography titled *Outlines of Military Geography* (Maguire 1899). Maguire described some good operational examples from history and introduced a broader ‘industrial age’ view of geography; that is, strategists should not just study the tactical terrain, the land battle, and what was over the next hill, but should also appreciate what sea and sea lines of communication were crucial to formulating a strategy that could be logistically supported.

Geographic information has long been a fundamental input to situational awareness (SA) which is recognised as critical for successful decision making in the military context. In military operations SA is a cognitive function derived from the COP, with geographies forming the basis of the picture and with other intelligence layered on top.

Since the invention of topographic maps in the eighteenth century, commanders have been layering operational and cultural information on a geographic base (Hindle 1998). Information such as the position of troops and the status of infrastructure together with the locations of population centres and other cultural information all contribute to the formation of the COP.

3 Present

Military operations in the information age have become more complex. Air, space and sub-surface operations add more layers to the traditional land and sea strategy.

Weapons have increased in reach and lethality, and can be delivered with increased precision and accuracy. Asymmetric and non-traditional threats add another cultural layer to the political landscape. Together these create complexities for military strategy that drives a need for deeper understanding of the geographies in which military forces operate. The modern military commander requires precise SA on a broad scale to achieve information superiority and respond rapidly to the changing battlespace.

The conversion of geographic data, commonly referred to as geospatial data, into useful information has been termed GEOINT.

As military operations have moved from small area strategic, operational and tactical processes, to global interactions and interdependencies, GEOINT informs the same processes but with an expanded range of spatial and temporal scales. Broad scale access to data and improved technological and human synthesis tools have facilitated a significant increase in the breadth and volume of GEOINT that forms the COP. The speed and interconnectedness of operations has benefited from the digital world and in particular the COP, which can be a dynamic display as a result of real time information. This capability was limited under the old paradigm of analogue maps.

To meet this challenge within the Australian context the Department of Defence established in 2000 the Defence Imagery and Geospatial Organisation (DIGO), an amalgamation of the Australian Imagery Organisation, the Directorate of Strategic Military Geographic Information and the Defence Topographic Agency. In 2013, DIGO was renamed to the Australian Geospatial-Intelligence Organisation.

> *Critical enablers, such as improved data access, flexible analytic tools and automated processing, will allow us to fully harness and exploit this integrated intelligence. We will be able to conduct complex GEOINT analysis in more efficient and effective ways.*
>
> *This will enable the production of richer GEOINT and deliver a more complete intelligence picture to our customers.* Australian Geospatial-Intelligence Organisation – Vision Statement 2016–2020 Mr. Neil Orme - Director.

The equivalent organisation in the United States of America is the National Geospatial Intelligence Agency. The commonly accepted definition of GEOINT is given by the US National Geospatial Intelligence Agency (2017):

Geospatial intelligence is the exploitation and analysis of imagery and geospatial information to describe, assess and visually depict physical features and geographically referenced activities on the Earth. GEOINT consists of imagery, imagery intelligence and geospatial information.

It can be argued that the definition above could be expanded to include 'the exploitation and analysis' of data from above and below the earth's surface to 'describe, assess and visually depict' the increasing array of geospatial and temporal information from multiple sources.

The scope and size of the COP has expanded, as regional powers gain ascendancy, and sensor capability and weapons reach continue to grow. The trend of technological advancement in both military operations and data gathering capability means the military geographer is now faced with the problem of, not a lack of data, but too much data. This is the 'big data conundrum' and the challenge posed to GEOINT practitioners is how to ingest, analyse, synthesise and interpret this data into meaningful and useable intelligence for the commander.

In the past GEOINT may have resulted from field surveys or the use of imagery collected from aircraft, over selected targets, depicted as printed photographs, analysed to produce specific intelligence to inform the COP. Today imagery is no longer constrained to the visible spectrum as contemporary sensors can collect data that is unable to be seen by the human eye and can be exploited with sophisticated software. Hyperspectral imaging has enabled the users to see through smoke, identify camouflage and detect changes not possible with the human eye. Furthermore, the use of Synthetic Aperture Radar sensors, that use radio waves that enable analysts to detect minute variations in ground surface elevation, provides relevant intelligence in detecting vehicle tracks across vast expanses of sandy deserts.

Motion imagery, often collected from remotely piloted aircraft, is another geospatial data source. It allows analysts to monitor patterns of activity and provide real-time intelligence to decision makers. Light detection and ranging, or LiDAR systems, use a pulsed laser to measure distance to objects to produce accurate elevation data. Using LiDAR, GEOINT analysts can accurately three-dimensionally map an urban landscape, the inside of buildings, or subterranean caves. LiDAR is also used to provide information on the structural complexity of woody vegetation (Lim et al. 2003). Advances in sensor technology has seen a move towards miniaturisation of sensors and a move away from air-breathing platforms to space-based platforms.

Geospatial data is not only collected by remotely piloted vehicles but is a fundamental input into their ability to operate. A ground control system for a remotely piloted vehicle requires terrain information to identify vertical obstructions for take-off and landing as well as foundation data for its own situational awareness. This need for GEOINT to operate future platforms is not limited to remotely piloted vehicles. As machines take on an increasing number of tasks in the decision-making chain, GEOINT will play a significant role in the development of customised machine learning to enable timely situational awareness at a tactical level. The move is away from platform centric warfare to network centric warfare where GEOINT becomes a critical input.

Today, GEOINT augments traditional geospatial data and imagery with data that has a location associated with it and a temporal footprint. The rise of social media applications has resulted in over two billion users worldwide (Statista 2017). Most of these applications and news feeds access a user's location which allows their posts, tweets and usage to be spatially identified. Analytical software can sift through vast volumes of this unstructured data, geotag it with a location and visualise it in a geographic and temporal context.

Personal mobile devices with their cameras, GPS and receiving and transmitting capability are increasingly, a rich source of GEOINT data. The tasks of collecting, fusing, deciphering and integrating the volume of data from these devices is immense and forms part of the "big data" issue. Temporal archives of GEOINT data allow the analyst to reach back into reference libraries to detect changes over time. The utility of such data is reliant upon quality data storage and accessibility, however the benefits of such data stores for counter terrorism operations are invaluable. It is also acknowledged that access to social media as a source of GEOINT is of concern in the public-private domain.

GEOINT is also crucial for the military planner in non-conflict environments, such as the case for humanitarian assistance and disaster relief (HADR). The temporal aspect of GEOINT allows for the analysts to determine what is normal and enables the analyst to highlight what has changed after an event. These processes are becoming more and more automated as technology advances. GEOINT has become a fundamental part of many HADR efforts where tasks such as identifying post event damage or the most appropriate locations to establish refugee settlements or communication facilities are able to be undertaken prior to a single person being placed on the ground.

4 Space

Space-based sensors now provide fundamental data to GEOINT in the areas of weather monitoring, communication and earth observation. The security of nation states and regional geopolitical alliances, modern military engagement and significant commercial and public services are all reliant on satellite-derived data and communication networks. For this reason, space-based assets are now a high priority for protection and must be described, assessed, and visually depicted by GEOINT scientists to inform an enhanced COP. As with land based assets, monitoring adversaries' space-based assets is now a vital input for knowing where your enemy is operating.

The reliance on space-based sensors was highlighted by Marcel Lettre (Acting Under Secretary of Defence and Intelligence) at the 2015 GEOINT Conference when he expressed concern about the growing threat from anti-satellite weapons being developed by China and Russia. Let's ponder that and the relevance of modern day military geography:

> *... ensuring that our future space-based intelligence, surveillance and reconnaissance (ISR) capabilities, including GEOINT collectors, are resilient against these threats and continue to provide support even in heavily contested conflicts, ...* (Lettre 2015)

Space Situational Awareness (SSA) refers to the ability to detect and track, understand and predict the physical location of natural and manmade objects in orbit around the earth. SSA has become a prominent concern for both military and commercial systems largely due to increasing military reliance on space-based assets. With the reduction in cost of building and launching smaller satellites into low earth orbits, space clutter has become a major issue. Anti-satellite testing by China in 2007 and the collision of a non-operational Russian satellite with an Iridium satellite in 2009 has also raised the urgent need for SSA.

US Deputy Defense Secretary Robert Work called for dramatic improvements in the GEOINT products available to the Department of Defense during his GEOINT 2015 keynote address. He discussed the increasing vulnerability of space-based systems and the need to make the US more resilient in the face of emerging, potentially adversarial powers such as China and Russia. Saying:

> *If we fail to do so, the implications for our national security will be quite profound. Our command and control would be significantly degraded. Our ability to detect and track adversary ballistic missile launches would suffer. The accuracy of our precision-guided munitions would be put into question. The satellite links that connect our aerial - unmanned aerial systems could be denied. We would lose much of the space-derived data that forms the foundation of nearly all of our intelligence products.* (Work 2015)

5 Future

Quantities of geospatial data are increasing exponentially as sensors improve and the scale of operations expand. Integrating all of this spatially enabled data to form a COP will rely on advanced analytics, high-end computing and highly trained analysts. The future will not see vast rooms of geospatial analysts pouring over imagery, as in Fig. 10.1, but instead a process chain that focusses on artificial intelligence (AI) and machine learning (ML) where the analyst is doing the coordination in a manner not dissimilar to a conductor and an orchestra. AI and ML will take over many of the mundane, repetitive processes currently undertaken by the analyst.

The GEOINT scientist of the future will still need to understand geography and all its interrelationships both on earth and in space, but their focus will not be on sensors, individual images or geotagging engines. It will be their ability to utilise artificial intelligence and machine learning to extract the information from big data that will be most valuable in the GEOINT process of developing SA and informing the COP.

Fig. 10.1 The 127th Command and Control Squadron is an example of current geospatial analysis https://commons.wikimedia.org/wiki/File:127th_Command_and_Control_Squadron_-_Distributed_Common_Ground_System.jpg

An understanding of geography and social data and how they relate in spatial-temporal domains is critical to the successful development of the algorithms behind the analytical engines. To achieve this end, it is essential for GEOINT scientists to have a clear picture of what is available and what is required. In addition, when delivering the resulting GEOINT to decision makers, it must be in a form which is easily interpreted: this domain is termed geospatial visualisation.

Just as cartography was a highly specialised skill required to create and communicate complex information to decision makers, todays' decision makers are expecting this complex synthesis of GEOINT data to be delivered ideally in real-time. This will rely on innovative spatial big data analytics and quality communications networks. The GEOINT scientist must deliver the information to digital displays ranging in size from a smart watch to a large wall size screen and be able to display multi-dimensional geospatial data across space and time. The design of the graphical user interface to facilitate interpretation of GEOINT will continue to increase in importance.

Future systems will operate as collections of various sensors, regardless of specific scale or geographies. They will be able to cross queue each other and to task and re-task other systems based on their role and the decisions required. There will be no boundary between space, air, land, water and subterranean. Artificial intelligence, automatic pattern and feature extraction and interpretation, automatic queuing and tasking of sensors will all be common place in the future. The GEOINT scientist will be the conductor, their expertise will be essential to understand the client's needs and how the various components interact with each other and the environment within geographic theatres.

GEOINT will be a fundamental input into fourth and fifth generation military hardware. GEOINT will feed the computers that allow these systems to perform effectively. Whether it is autonomous vehicles or Counter Unmanned Aerial Vehicles, a digital roadmap will be required for this technology to operate. The push to use fifth generation military equipment is resulting in a refreshed interest in GEOINT and the core capabilities and skills required to address the big data geospatial issues. The push for science, technology, engineering and mathematics (STEM) skills in Australia is directly related to this current and future skill gap.

Dr. Michael Hauck, (Hauck 2017) a geospatial consultant and former director of the American Society for Photogrammetry and Remote Sensing, at the GEOINT 2017 (San Antonio) stated:

> *We have more than one specialization. Specialized LiDAR technology and analysis. Specialized radar. Specialized, specialized, specialized. And machines now are becoming very specialized.*

And then asked the question

> *But to solve real-world intelligence problems, is it specialization … or is it overall understanding of how this fits together that is going to be critical?*

6 Conclusions

The question being asked by the commander today and in the future, will remain, "What is over the hill, underneath the hill and above the hill?" The hill will only be figurative as it may refer to a continent and the picture that is presented to the commander may be a moving display on his mobile phone or a hologram in his command centre. Answers to this question and others will require the continued development and application of advanced technologies to synthesise massive amounts of multi-dimensional geographic reference data and to present the results intelligibly on all manner of devices in real time. The GEOINT scientist will play a pivotal role in strategic and military geography long into the future.

References

Clausewitz, C., Howard, M., Paret, P., & Brodie, B. (1984). *On War*. Princeton, New Jersey: Princeton University Press.

Hauck, M. (2017). GEOINT 2017 presentation. GEOINT 2017 Symposium, San Antonio, Texas, USA.

Hindle, P. (1998). *Maps for historians* (pp. 114–115). West Sussex, UK: Phillimore & Co.

Lettre, M. (2015). GEOINT 2015 Keynote address. GEOINT 2015 Symposium, Washington DC, USA. Online at: http://geointv.com/archive/marcel-lettre/ [Accessed 14 July 2017].

Lim, K., Treitz, P., Wulder, M., St-Onge, B., & Flood, M. (2003). LiDAR remote sensing of forest structure. *Progress in Physical Geography, 27*(1), 88–106.

Maguire, T. M. (1899). *Outlines of military geography*. Cambridge, MA: Cambridge University Press.
National Geospatial Intelligence Agency. (2017). Mission Statement. Online at: https://www.nga.mil/ProductsServices/Pages/default.aspx [Accessed 14 July 2017].
Statista. (2017). Number of social media users worldwide from 2010 to 2020 (in billions). Online at: http://www.statista.com [Accessed 14 June 2017].
Sun-Tzu, & Griffith, S. (1964). *The Art of War*. Oxford, UK: Clarendon Press.
Work, R. (2015). GEOINT 2015 Keynote address. GEOINT 2015 Symposium, Washington, DC, USA. Online at: http://geointv.com/archive/robert-work/ [Accessed 14 July 2017].

Chapter 11
Characterising the Environmental Values of the National Defence Estate, with Emphasis on Native Vegetation

Richard Thackway and Frederick Ford

1 Introduction

Australia's Defence estate (Fig. 11.1), currently covers around 3.4 million hectares (Department of Defence 2009) (0.45% of Australia's land area), and this includes relatively undeveloped training areas, bases, depots and a variety of other functional purposes (Fig. 11.1). Training areas represent the majority of the land area, either as stand-alone properties with limited facilities, or smaller training areas adjoining bases in peri-urban or rural areas. Most of the estate retains its native vegetation extent, which serves various functions including cover for Defence training, the protection of ecosystem services and the maintenance of biodiversity (Appendix Table 11.5).

Defence lands internationally are generally located in natural and underdeveloped landscapes and represent a distinct and important land use that often requires exclusion of general access and strict control over activities and uses of the landscape. A recent international inventory (Zentelis and Lindenmayer 2015) showed the potential importance of Defence training areas as a part of conservation, although they are not managed exclusively to secure long-term conservation. In Australia, as elsewhere, Defence properties include habitats for key threatened species and communities (Bell 1985; Zentelis and Lindenmayer 2015), comprise major components of landscape connectivity (Fitzsimons et al. 2013) and contribute to the provision of

R. Thackway (✉)
UNSW Canberra at the Australian Defence Force Academy, Canberra, ACT, Australia

F. Ford
Brindabella Business Park, Canberra, ACT, Australia
e-mail: frederick.ford@defence.gov.au

S. Pearson et al. (eds.), *Australian Contributions to Strategic and Military Geography*, Advances in Military Geosciences,
https://doi.org/10.1007/978-3-319-73408-8_11

Fig. 11.1 Distribution of major Defence facilities in Australia

ecosystem services within and beyond their boundaries including biodiversity, soil, water and carbon sequestration (Yapp et al. 2010). Core biodiversity values may also represent culturally valued species or landscapes within all parts of the community, but particularly within Indigenous peoples' communities (Natural Resource Management Ministerial Council 2010). The Defence estate is also an important part of Australia's environmental accounts (Bureau of Meteorology 2013) and it is vital knowing what contribution various natural landscapes (assets) and their condition classes, such as the Defence estate, make to Australia's environmental accounts (Sbrocchi et al. 2015).

As one of Australia's largest land managers, Defence is sometimes challenged by the need to balance the subsequent development of these ecosystem services and values against maintaining the core function of the Defence estate: to house and sustain Defence capabilities and provide adequate training opportunities. Bioregions that are dominated by native vegetation are increasingly being recognized and managed to maintain biodiversity and build resilience (Natural Resource Management Ministerial Council 2010). Reporting on the protection of native vegetation type, extent and condition is a key environmental outcome (COAG Standing Council on Environment and Water 2012). Defence recognises that the requirements for environmental management extend beyond simply allowing vegetation to improve in condition and extent. The Defence Environmental Strategy 2016–2036 highlights the requirement to manage the potential risks posed to neighbours and the wider community that are associated with large areas of vegetation, including bushfire hazard, and the harbouring of pests, weeds and disease (Department of Defence 2016). Meeting these challenges requires access to up-to-date nationally consistent relevant environmental datasets which describe ecosystem type, extent and condition, and a capacity to monitor changes and trends in resource condition against benchmark conditions (Thackway and Freudenberger 2016). In Australia, vegetation is a widely accepted descriptor for ecosystems in many environmental assessments (Yapp and Thackway 2015).

The location of Defence properties (Fig. 11.1) is an artefact of historical accident along with past and current strategic requirements. Concentrations of Defence facilities in southern Australian landscapes have gradually been replaced following recognition of the requirement for a northward facing force posture. Increasingly large training areas are required to accommodate modern weapons systems and their safety templates, and these areas must span strategically relevant environments and be relatively accessible to force elements that require them for training. This has resulted in a continent-wide distribution of properties that area-wise is dominated by northern Australian training areas (Fig. 11.1), but which still retains a significant presence in the intensively managed southern and eastern latitudes where European settlement has had comparatively greater impact (Thackway and Lesslie 2008). In these more intensively managed landscapes, that is where landscapes are more highly fragmented and modified, individual Defence training areas tend to be relatively large areas, and are internally less fragmented and modified compared to their surrounding landscape. This is discussed below. It is worth noting that almost all Defence facilities were previously managed for other resource management purposes e.g. forestry and grazing (Thackway and Pearson 2017). In addition, prior to acquisition and management for Defence purposes, the previous land uses variously modified the function, structure and composition of the plant communities. Over time Defence management has also variously affected the condition of these properties (Thackway and Pearson 2017).

In this chapter, we characterise Australia's Defence estate using several national vegetation datasets that describe the type, extent and condition of native vegetation. Within the more fragmented and modified landscapes of eastern mainland Australia (Thackway and Lesslie 2008), lie numerous defence training areas, which because of their proximity to major population centres are frequently used for Defence purposes. These training areas also span the latitudinal range of eastern Australia and represent the variety of Defence land use types that are readily accessible to Defence personnel along the eastern seaboard. To highlight the differences in the ecological integrity and transformation of landscapes between individual Defence facilities and the surrounding landscapes in eastern mainland Australia, we selected eight training areas (Fig. 11.2).

We analyse and discuss these eight case studies, noting how the management of Defence estate generally maintains the extent and condition of native vegetation. We discuss the role Defence land may play in safeguarding conservation and

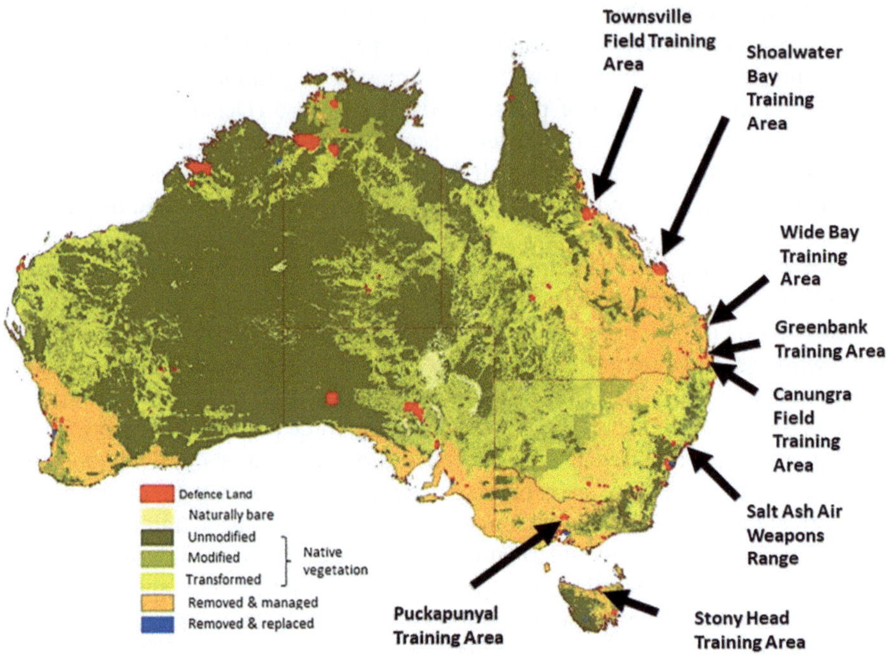

Fig. 11.2 Distribution of the eight select DoD estates. Background spatial dataset shows the VAST 2010 dataset

heritage values, and ecosystem resilience. Finally, we discuss the importance of using consistent national environmental datasets to characterise the Defence estate. The role of such information has the benefit of informing decisions of Defence land planners and policy managers to affect changes and trends in vegetation condition (function, structure and composition) in accordance with the strategic and operational objectives for each Defence facility, and to achieve and maintain each training area as fit for purpose.

2 Method

2.1 Characterizing the Defence Estate's Environmental Profile

The extent of the Defence estate was determined based on 2015 data held by the Department of Defence (Fig. 11.1). This extent is not fixed, and tens of thousands of hectares of new land has since been purchased by Defence for inclusion in training areas, while smaller redundant sites have been divested.

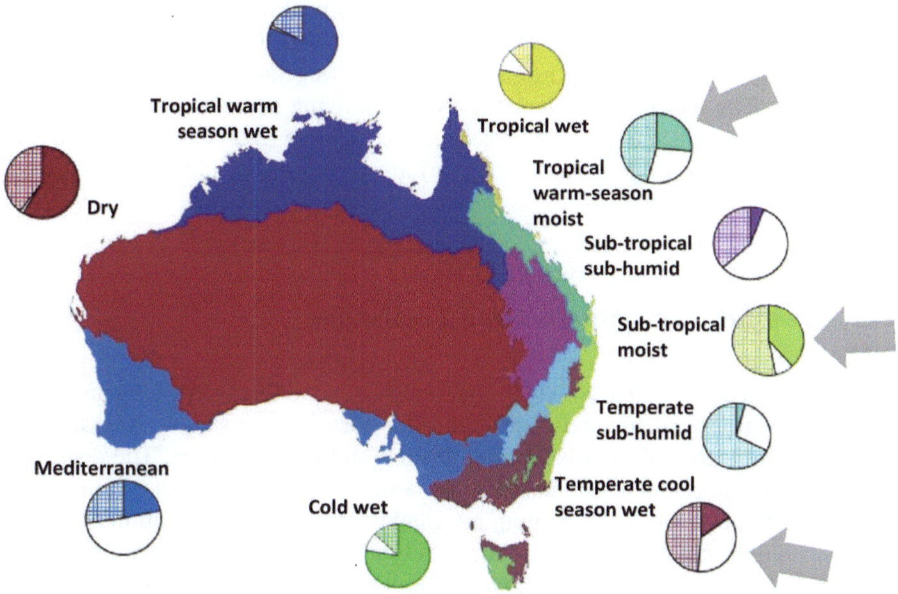

Fig. 11.3 Ten agro-ecological regions showing the three regions within which the eight Defence facilities were selected. The pie charts show the degree of modification (condition classes) of native vegetation within each region. Description of pie charts: Solid fill: Naturally bare and Unmodified vegetation, varying condition (VAST 0–1), Grid: Modified vegetation (VAST 2–3) and Open: Native vegetation replaced and Removed and replaced (VAST 5–6)

We used a geospatial analysis to characterise Australia's Defence estate by intersecting the extents of Defence facilities (Fig. 11.1) using four widely recognised environmental national geo-spatial datasets:

1. Major Vegetation Groups (Department of Environment and Energy 2016a and b);
2. agro-climatic regions (Hobbs and McIntyre 2005; Thackway and Freudenberger 2016, Fig. 11.3);
3. Vegetation condition (Lesslie et al. 2010); and
4. Landscape Alteration Levels (Mutendeudzi and Thackway 2010).

These four datasets describe the distribution, type, extent and condition of native vegetation. These four datasets have been developed by the Australian Government and its partner state and territory land management and research agencies, and are widely used by the Australian Government (e.g. Mutendeudzi and Stafford-Bell 2011). Lists of the classes in the four datasets and the relative areas of each class across Australia are presented in the Appendix: agro-climatic regions (Table 11.3); Major Vegetation Groups (Table 11.4); condition classes (Table 11.5) and Landscape Alteration Levels (Table 11.6).

The Landscape Alteration Levels dataset was derived by using a compositional metric available in FRAGSTATS (McGarigal et al. 2002) Version 3.3 software program which measures the proportion of the various vegetation condition classes or patch types. The metric was calculated using a moving square window approach, which can be imagined as a window of fixed size passing over every grid cell of the input dataset. Area proportions for the various vegetation condition classes were calculated during the process.

2.2 Remnants in Transforming Landscape

We investigated the degree to which the Defence estate represents remnants in an increasingly transformed landscape along the eastern seaboard of Australia, by assessing the degree of landscape modification and fragmentation. The degree of landscape transformation was analysed by extending a 10 kilometre buffer area around the boundary of each training area and comparing the vegetation condition found within the eight training areas to the neighbouring landscapes, within the buffer area. Selection of a 10 km buffer was considered sufficient to encapsulate change in the area surrounding each Defence estate and was adequate relative to the size of each Defence estate. Spatial analyses were undertaken as zonal analyses in the Spatial Analyst extension of ArcGIS 10. Zonal analyses were used because these tools allow computations on all cells that belong to the same condition class, within and outside the zone. The input condition dataset was a raster (Fig. 11.2).

The degree of landscape transformation shown in Fig. 11.2 is based on the Vegetation Assets, States and Transitions (VAST) framework (Thackway and Lesslie 2006). The VAST framework describes and accounts for changes in the extent and condition of native vegetation (note these two parameters may converge spatially in some circumstances). VAST makes explicit the links between land management and vegetation modification.

We examined the degree of landscape transformation in the context of the ten agro-climatic regions (Fig. 11.3). The degree to which the native vegetation in each agro-climatic region has been transformed by land use and management regimes is depicted as pie charts using aggregations of condition classes (Appendix Table 11.5). Eight training areas were selected primarily because of their importance as high-use training areas and/or span across the eastern latitudinal range of eastern Australia. These training areas reflect the variety of Defence land use types that are readily accessible to Defence personnel along the eastern seaboard (Table 11.1). This table also provides the corresponding agro-climatic region. The location of the selected training areas is shown in Fig. 11.2.

It should be noted that none of the selected Defence training areas are found in the following six agro-climatic classes: Dry, Mediterranean, Cold wet, Temperate sub-humid, Sub-tropical sub-humid, and Tropical wet.

Table 11.1 Description of the eight Defence training areas

Case study training area	Year established	Area (ha)	Current key use of training area	Locality and jurisdiction	Agro-climatic class
1. Townsville field training area	1967	183,514	Live-fire military range for the Australian defence force	Hervey range, Queensland (Qld)	Tropical warm-season moist
2. Shoalwater Bay field training area	1965	264,144	Amphibious operations training ground and combines blue-water naval forces with comparable air and land assets	Ogmore, Qld	Tropical warm-season moist
3. Greenbank training area	1940s	4505	Live weapons and firing range	Larapinta, Qld	Sub-tropical moist
4. Canungra field training area	1940s	5836	Land warfare Centre	Canungra, Qld	Sub-tropical moist
5. Wide Bay training area	Pre 1980s	19,127	Combat team training	Tin can bay, Qld	Sub-tropical moist
6. Salt Ash air weapons range	Mid 1950s	2754	Air to ground gunnery and bombing facility	Medowie, New South Wales (NSW)	Sub-tropical moist
7. Puckapunyal military area	1939	42,442	Training activities involving the use, distribution and disposal of ammunition, and maintenance of vehicles and equipment	Seymour, Victoria (Vic)	Temperate cool season wet
8. Stony Head training area	Pre 1960s	5567	Various firing ranges, and the disposal of ammunition in dedicated disposal area	Beechford, Tasmania (Tas)	Temperate cool season wet

3 Results: Characterising the Defence Estate

3.1 Agro-Climatic Regions

The total area of the Defence estate is around 3,492,000 ha or 0.45% of the land area of Australia (Appendix Table 11.3). The estate is distributed in all of Australia's 10 agro-climatic regions, except the Cold wet (i.e. alpine and sub-alpine areas) (Fig. 11.3, Appendix Table 11.3). The absence of any Defence estates in the Cold wet agro-climatic region reflects that this agro-climate type is not regarded as a strategic priority for the Australian Defence forces.

Over 60% of all Defence facilities are located in only three agro-climatic regions: Temperate cool-season wet, Sub-tropical moist and Tropical warm season moist

(Appendix Table 11.3). The largest areas of the Defence estate (i.e. larger than 500,000 ha) are found in three agro-climatic regions; Tropical warm season wet (1,783,448 ha or 2.16% of this region's total area), Dry (907,841 ha or 0.18% of its total area) and Tropical warm season moist (524,446 ha or 2.15% of its total area) (Appendix Table 11.3). The smallest area of the Defence estate (i.e. smaller than 10,000 ha) is found in two agro-climate regions; Temperate - Sub-humid (6194 ha or 0.04% of its total area) and Sub-tropical Sub-humid (1669 ha or circa 0.00% of its total area).

3.2 *Vegetation Types*

Our assessment shows that within the Defence estate there is a lot of vegetation diversity and this is a good thing as it gives the Australian Defence Force access to a wide range of vegetation types for training in different classes of cover densities, height and shapes of vegetation (physiognomy) as well as trafficability[1] and line of site.

The Defence estate in Australia includes 14 of the 33 national vegetation types (i.e. Major Vegetation Groups (MVG) Appendix Table 11.4). Examining the types of vegetation found in the estate, shows that there is a high proportion of tree dominated types (woodland and open woodlands), compared to shrub and grass (or grass-like) dominated types. The five largest areal extents of these vegetation types found in the Defence estate (i.e. larger than 200,000 ha) include class 5 Eucalypt Woodlands (1,081,521 ha or 1.28% of its national extent), Hummock Grasslands (375,554 ha or 0.27% of its national extent), class 13 Acacia Open Woodlands (344,063 ha or 0.90% of its national extent), class 11 Eucalypt Open Woodlands (260,311 ha or 0.56% of its national extent), and class 12 Tropical Eucalypt Woodlands/Grasslands (215,152 ha or 1.59% of its national extent). All of these MVGs have widely-spaced trees, enabling substantial understories. While these areas are large within the estate, it is however helpful to place these types within a national context. For example, compare the national extent of class 5 Eucalypt Woodlands, which comprises 84,528,297 ha or 10.99% of the land area of Australia, relative to its extent within the Defence estate (1,081,521 ha) (Appendix Table 11.4).

The remaining 27 national vegetation types found in the Defence estate only occur as very small relative extents (i.e. less than 1% of their broader national extents) (Appendix Table 11.4). Of these 27 types, class 25 Cleared, Non-Native Vegetation, Buildings, comprising 97,752 ha or 0.09% of its national extent, occurs in the Defence estate. An additional class 99 Unknown/No Data, comprises 1,350,186 ha nationally and of this area, only 6588 ha or 0.49% of its national extent, occurs in the Defence estate.

[1] The ability of a particular vehicle to traverse a specified terrain.

3.3 Vegetation Condition

The condition of Australia's vegetation is presented in six classes (Appendix Table 11.5) that relate to three broad types of vegetative cover: Naturally bare (class 0); Native vegetation (Unmodified, Modified and Transformed, classes 1, 2 and 3 respectively), and Non-native vegetation (class 4 Adventive and class 5 Removed and managed vegetation). Where the native vegetation has been removed and replaced with non-vegetated land cover types e.g. water and infrastructure, this is class 6. More information about these classes and the VAST framework are presented in Thackway and Lesslie (2006; 2008).

VAST is a simple communication and reporting tool designed to assist in describing and accounting for human-induced modification of vegetation. VAST class I, Unmodified, provides the benchmark for each vegetation association based on structure, composition and current regenerative capacity (ecological function). Benchmarks are based on the best understanding of pre-European conditions (sometimes called 'fully natural'). In Fig. 11.2, relative change in condition from this benchmark is assessed for each raster cell, relative to class 1, i.e. Unmodified.

The two largest areal extents of these vegetation condition classes found in the Defence estate (i.e. larger than 500,000 ha) include Residual or Unmodified (class 1) (2,158,121 ha or 0.64% of its national extent), Modified (class 2) (836,938 ha or 0.41% of its national extent).

Compared to these three larger classes of managed native vegetation, Class 5, Replaced and managed for intensive use production, is relative small in extent (96,355 ha or 0.09% of its national extent). Three other classes occur as minor components of the Defence estate class 0 (Naturally bare 10,166 ha or 0.09% of its national extent) and Class 6, Replaced with man-made structures/infrastructure, (8746 ha or 0.92% of its national extent). Class 4 is not mapped in the national condition dataset and so is absent from this assessment of the Defence estate.

3.4 Landscape Alteration Levels

Landscape Alteration Levels (LAL) describe the extent to which landscapes are both fragmented and modified. Landscapes are classified into four classes (intact; variegated; fragmented; and relictual) based on the degree to which the native vegetation has been fragmented and modified (Table 11.2). These four levels of landscape alteration described by McIntyre and Hobbs (1999, refer to Fig. 5) are widely known and understood across Australia's landscapes and assist conservation biologists and natural resource managers to address the full spectrum of human landscape impacts.

The degree of landscape alteration in Australia is presented in four classes (Appendix Table 11.6). The two largest areal extents of these landscape alteration classes found in the Defence estate (larger than 1 million ha) include Intact >90

Table 11.2 Estimated proportions of condition classes (McIntyre and Hobbs 1999, Fig. 5) within each Landscape Alteration Level. Based on Mutendeudzi and Thackway (2010, Table 11.3)

	Naturally bare areas and Native vegetation				Non-native and Removed native vegetation
Landscape alteration levels	Naturally bare (class 0) unmodified (class 1)	Modified and retained (class 2)	Highly modified (class 3)	Mean sum of modification states (VAST classes 0–3)	Removed and managed (class 5) and removed and replaced (class 6)
Intact	>80%	>5%	>1%	>90% native vegetation retained	<10%
Variegated	>20%	>10%	>10%	60–90% native vegetation retained	10–40%
Fragmented	>1%	>1%	>1%	10–60% native vegetation retained	40–90%
Relictual	Remaining unallocated data	Remaining unallocated data	Remaining unallocated data	< 10% native vegetation retained	>90%

retained (1,864,540 ha or 0.53% of its national extent) and Variegated >60–90 retained (1,515,306 ha or 0.47% of its national extent).

The other two classes of Landscape Alteration Levels are very small by comparison, Fragmented 10–60 retained (79,001 ha or 0.14% of its national extent) and Relictual <10 retained (12,215 or 0.03% of its national extent).

3.5 *Remnants in Transforming Landscapes*

Figure 11.4 (a–g) takes a closer look at the degree of modification of native vegetation and the patterns of fragmentation within and surrounding eight major Defence training areas, situated along the eastern seaboard from north Queensland to northern Tasmania.

A comparison between the areas of native vegetation (modification i.e. class 1 Unmodified, class 2 Modified and class 3 Transformed) and non-native vegetation (fragmentation i.e. class 5 Removed and managed and class 6 Removed and replaced), made on Fig. 11.4a–g, show that there is a larger extent of non-native vegetation in the surrounding buffered areas than in the Defence training areas. Where native vegetation is present both inside and outside the training areas, the class 1 Unmodified is higher inside the training areas, showing higher integrity of the native vegetation than that in the surrounding 10 km buffer zone outside each

a

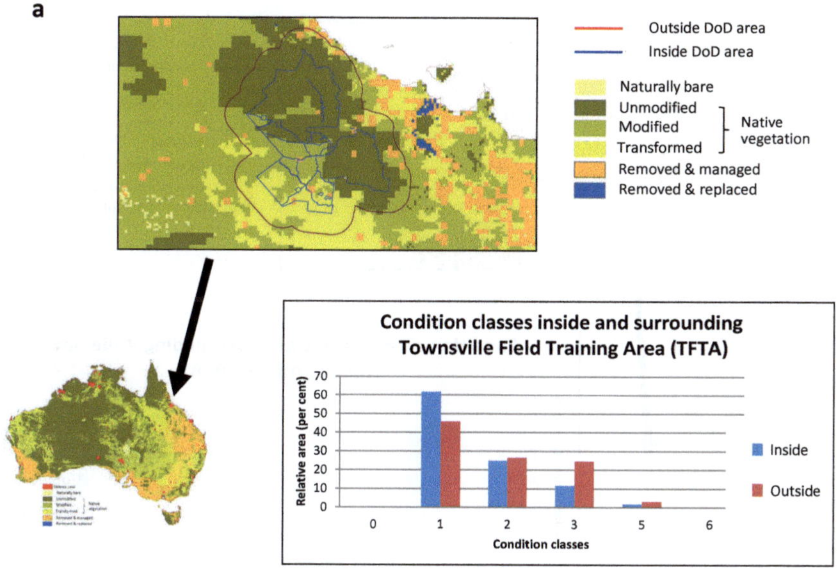

Fig. 11.4a Condition classes inside and surrounding Townsville Field Training Area (TFTA), Qld

b

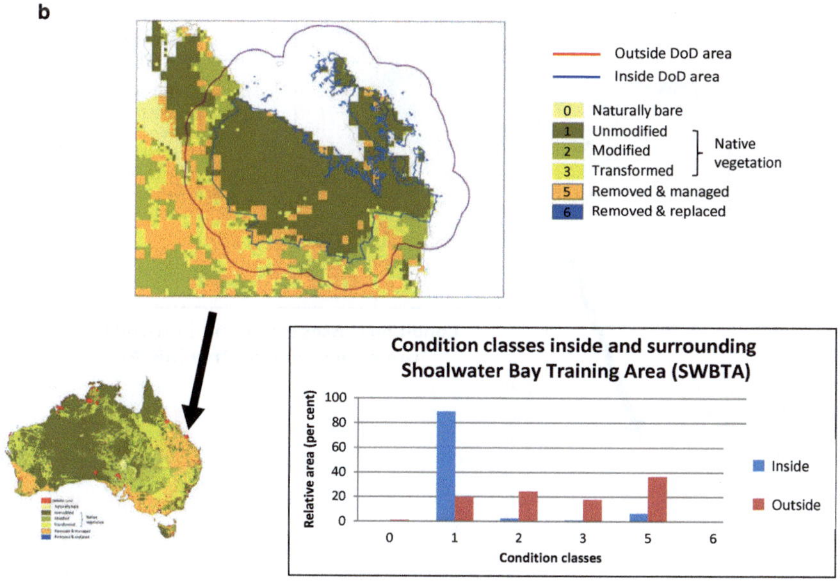

Fig. 11.4b Condition classes inside and surrounding Shoalwater Bay Training Area (SWBTA), Qld

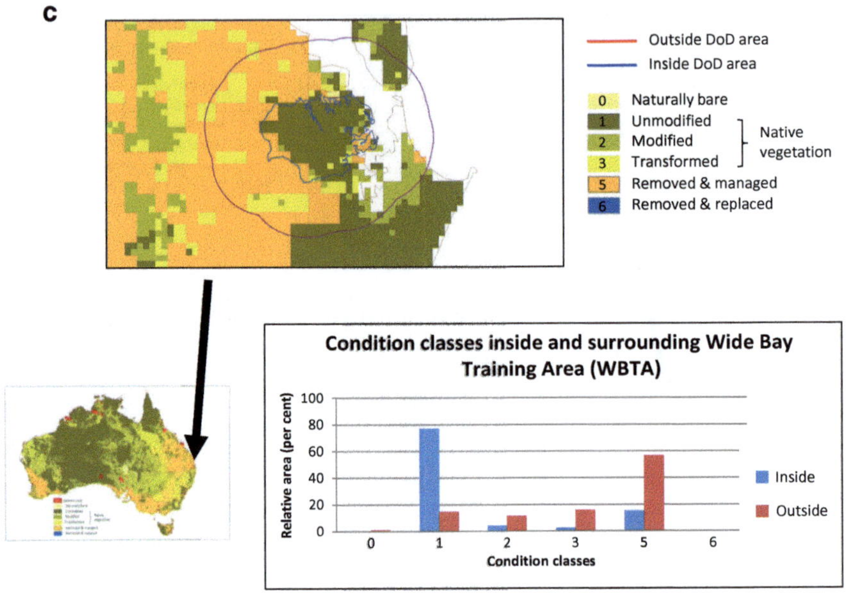

Fig. 11.4c Condition classes inside and surrounding Wide Bay Training Area (WBTA), Qld

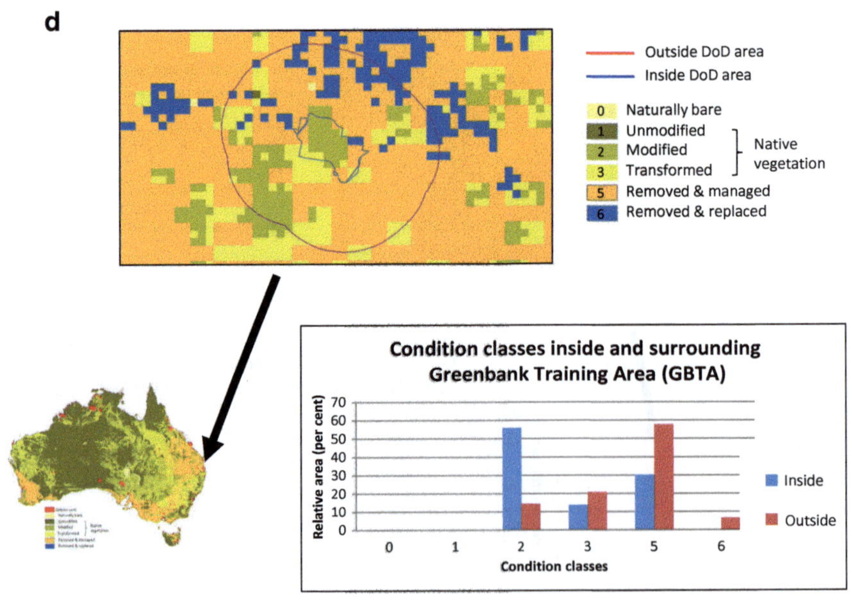

Fig. 11.4d Condition classes inside and surrounding Greenbank Training Area (GBTA), Qld

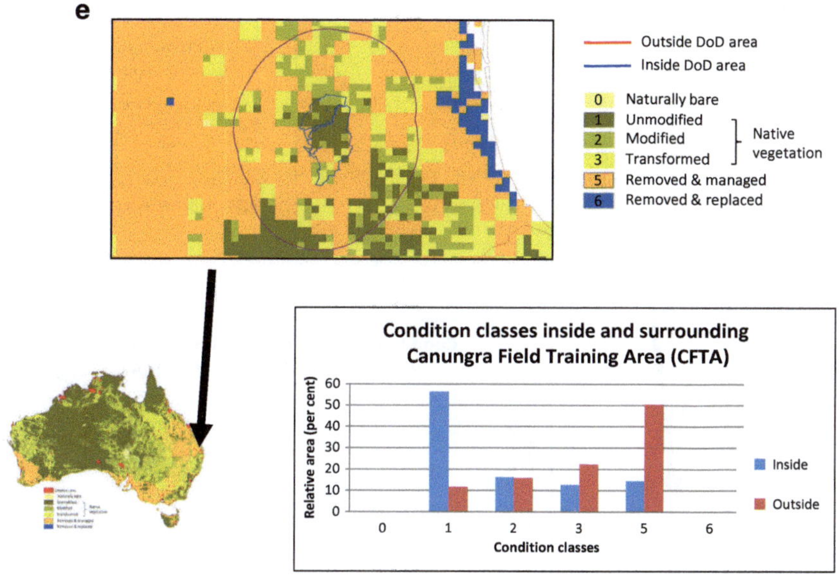

Fig. 11.4e Condition classes inside and surrounding Canungra Field Training Area (CFTA), Qld

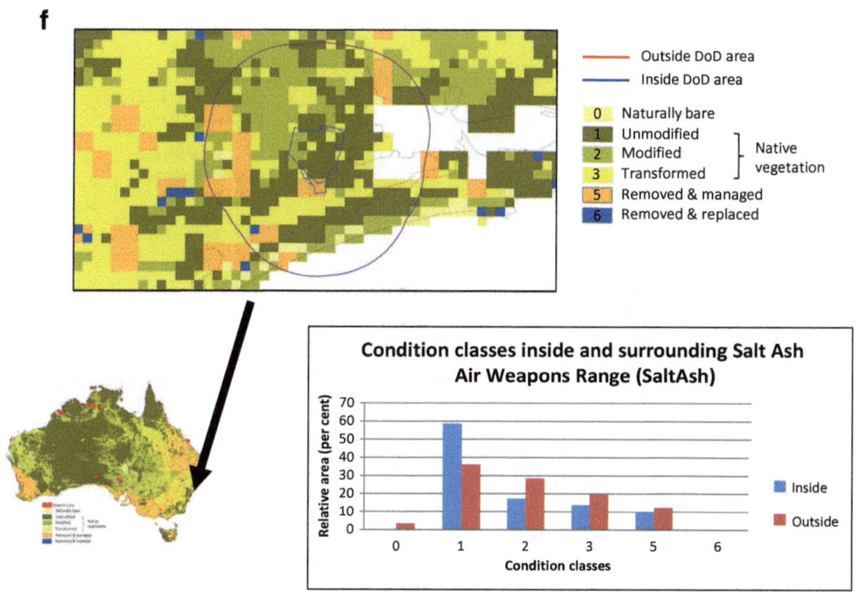

Fig. 11.4f Condition classes inside and surrounding Salt Ash Air Weapons Range (SaltAsh), NSW

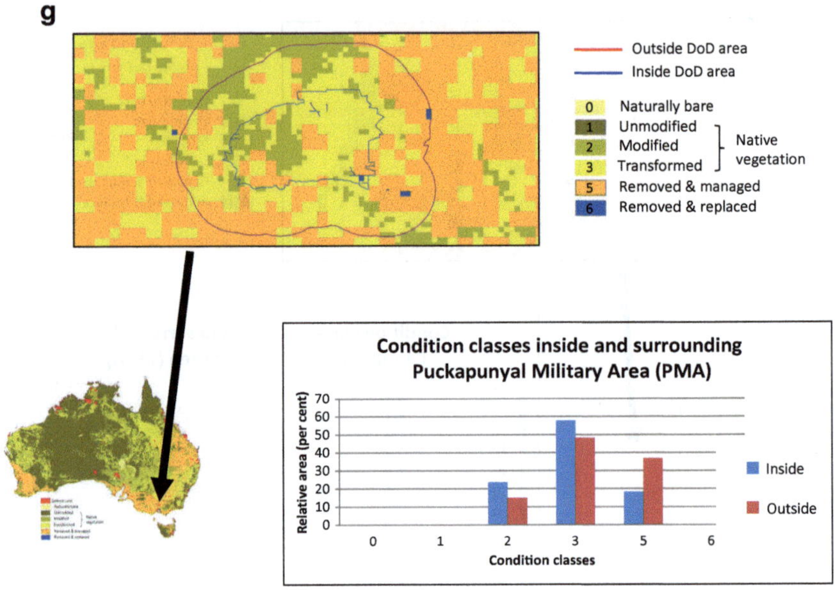

Fig. 11.4g Condition classes inside and surrounding Puckapunyal Military Area (PMA), Vic

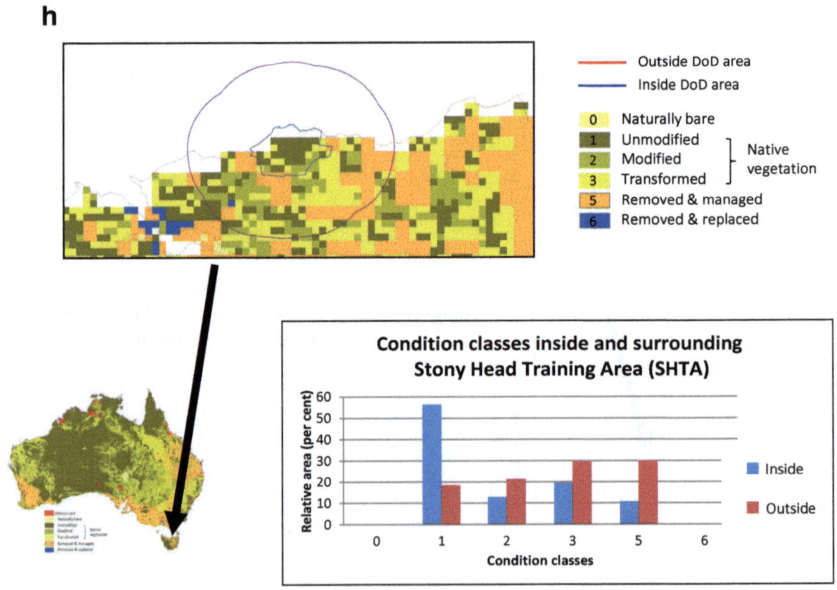

Fig. 11.4h Condition classes inside and surrounding Stony Head Training Area (SHTA), Tas

training area. The exceptions are Greenbank (Fig. 11.4e) and Puckapunyal (Fig. 11.4g) both of which do not contain class 1 Unmodified vegetation condition inside or outside the training areas. Looking further at these two training areas, Greenbank (Fig. 11.4e) and Puckapunyal (Fig. 11.4g), both record higher levels of class 2 Modified vegetation within these two areas compared to outside.

Looking at class 1 Unmodified, the highest level of integrity of native vegetation—inside and outside training areas—shows that Townsville (Fig. 11.4a), Salt Ash (Fig. 11.4f) and Stony Head (Fig. 11.4g) all have the highest levels of integrity of native vegetation inside and outside the training area across the three classes of native vegetation: class 1 Unmodified, class 2 Modified and class 3 Transformed. In contrast, there are obvious differences in the highest level of integrity of native vegetation between extents inside and outside some training areas e.g. Shoalwater Bay (Fig. 11.4b), Wide Bay (Fig. 11.4c), Canungra (Fig. 11.4d), with all three showing relatively smaller areas of class 1 outside the training areas.

Looking at class 3 Transformed, all training areas, except for Puckapunyal (Fig. 11.4g), show smaller relative areas of this class within the training areas, than the areas of this class outside the training areas. This shows the management regime of Puckapunyal, which is maintained for high trafficability, is modified pastures compared to the other training areas. Similarly, looking at class 5 Removed and managed, all training areas show smaller relative areas of this class within the training area, than the areas of this class outside the training area (Fig. 11.4a–h).

In summary, these results report the degree of modification of native vegetation (Fig. 11.4a–g) and the patterns of fragmentation, shown on map inserts inside and outside these training areas, record that the landscapes within training areas are less modified and less fragmented than the surrounding 10 km buffered areas.

4 Discussion

This chapter has described the extent and distribution and composition of the Defence estate in Australia using several consistent and widely used national datasets that describe the type, extent and condition of native vegetation and Australia's agro-climatic regions.

A feature of a national assessment is the very low areas (ha) and percentages of vegetation types, and their extents and condition, referred to above. If the same analyses were done using local and regional scale datasets, this would give the reader a different appreciation of the role and contribution of individual training areas (e.g. Bell 1985; Bowett et al. 2012; PGM Environment and Eco Logical Australia 2014). Such local and regional assessments would highlight the importance of individual Defence estates in terms of the number of species present, and in the diversity of species assemblages present. At these finer scales, the relative contributions of individual Defence estates would be seen in terms of maintaining and improving the connectivity of natural areas in landscapes managed using cooperative public-private partnerships. In addition, local and regional scale assessments would highlight the role and contribution of individual Defence estates in maintaining and protecting habitats for selected threatened species and threatened ecological communities. At local and regional scales, time series datasets could be used to show how land use pressures are impacting on the management of some defence estates (e.g. RAAF Base Amberley in Thackway and Pearson 2017).

Australia's natural landscapes have been transformed, i.e. modified and fragmented, wherever public and private drivers have promoted and enabled the development of landscapes, from local resource harvesting to mandated landscape-scale clearing. These effects are manifest at local, regional and national levels (Thackway and Lesslie 2008; Yapp and Thackway 2015) whether through deliberate or inadvertent changes wrought by prolonged management regimes that promote the modification and fragmentation of native vegetation. The relatively higher degree of modification of vegetation outside the Defence estate when compared to within it is particularly obvious along the eastern seaboard of Australia. The Defence estate can therefore play an important role in linking and supplementing other relatively contiguous extents of remnant native vegetation that are the focus of landscape connectivity initiatives (Fitzsimons et al. 2013). Further geospatial evaluation of the four datasets would allow for a GIS map/product to be prepared that can highlight contiguous extent at a regional scale.

The Defence estate generally maintains, protects and enhances native vegetation condition alongside supporting core Defence purposes, particular broadacre training requirements. Because most of the Defence training estate is not open to the public, the effects of Defence management on vegetation are therefore largely unseen and unknown. This contrasts with landscapes outside training areas which are owned by a multitude of land managers from large grazing leases to small rural holdings. Each has different ideals and aspirations, generally involving "improving" land condition for production or amenity, commonly through changing soil chemistry and composition, modifying existing vegetation or re-planting ornamental and exotic species. In such intensively managed landscapes, the landscape matrix continues to be developed and transformed as land managers change land-uses and land management regimes in response to various social, economic and environmental drivers.

The evidence presented above, showing gross changes in vegetation condition inside and outside training areas, highlights the obvious contribution that Defence estates make to maintaining the ecological integrity of landscapes across Australia. Additional work to prepare regional assessments using temporal land use and native vegetation datasets, at local and regional scales, would further demonstrate the contribution that Defence estates make to supporting public-private connectivity of natural areas, and maintaining natural areas as habitats for threatened species and threatened ecological communities (e.g. Bell 1985; Bowett et al. 2012; PGM Environment and Eco Logical Australia 2014).

Our results show that the total area of the Defence estate is more than 3,491,000 ha or 0.45% of the land area of Australia. The Defence estate is distributed in all of Australia's 10 agro-climatic regions, except the Cold wet (i.e. alpine and sub-alpine areas). The largest areas of the Defence estate are found in three northern agro-climate regions; Tropical warm season wet, Dry, and Tropical warm season moist. Conversely, of the total of 683 Defence facilities, 427 or 63% are found in three southern agro-climatic regions; Temperate cool-season wet, Sub-tropical moist and Tropical warm season moist (Fig. 11.2).

The Defence estate includes 14 of the 33 national vegetation types that describe the vegetation of Australia (i.e. Major Vegetation Groups). Examining the types of vegetation found in the estate shows that there is a high proportion of tree dominated types (woodlands and open woodlands), compared to shrub and grass like dominated types. Non-native vegetation cover types, while present in the Defence estate, are relatively small by comparison. The five largest areal extents of the major vegetation groups found in the Defence estate include Eucalypt Woodlands, Hummock Grasslands, Acacia Open Woodlands, Eucalypt Open Woodlands, and Tropical Eucalypt Woodlands/Grasslands. While these areas are large within the estate, they are widespread types and Defence manages relatively small areas of their total national extents.

The Defence estate is primarily characterised by two vegetation condition classes: Residual or Unmodified and Modified. We found that at a broader landscape scale, Defence has a preference to acquire new training areas where there is a predominance of Intact (> 90 retained native vegetation) and Variegated (> 60–90 retained native vegetation) Landscape Alteration Levels, compared to more highly fragmented and transformed landscapes i.e. Fragmented and Relictual landscapes.

The above results show that Defence management generally maintains and protects the natural values within their estate, while retaining sufficient land area to support ongoing Defence requirements. In the context of developing environmental-economic accounts, the Defence estate is also an important part of Australia's environmental accounts (Bureau of Meteorology 2013) and it is vital knowing what contribution various natural landscapes (assets) and their condition classes, such as the Defence estate, make to Australia's environmental accounts (Sbrocchi et al. 2015). We have shown that these national assets and their condition have been minimally modified and, as such, would be delivering multiple public services and benefits including clean water, healthy soils, landscape connectivity and threatened species habitats. These kinds of benefits were suggested previously from the Defence estate (Bell 1985; Yapp et al. 2010; Zentelis and Lindenmayer 2015). We have shown by inference, using the condition dataset based on the VAST framework, that Defence estate managers deploy conservative land management regimes which effectively maintain, protect and enhance the integrity of the native vegetation.

The value of the above results lies in the use of nationally consistent and authoritative datasets that are sponsored by the Australian Government (e.g. Mutendeudzi and Stafford-Bell 2011). We have combined these datasets using a standardised assessment framework, benchmarked to assess changes and trends at landscape scales relative to a reference state. Such an approach can assist other national monitoring, evaluation and reporting endeavours (e.g. State of the Environment Report 2011; MPIG 2013 Thackway and Specht 2015). Further work could be done at local and regional scales by analysing and reporting multi-temporal remotely sensed satellite image archives, combined with monitoring, evaluation and reporting of on-ground control points (Thackway and Pearson 2017 this volume). There would also be value in using relevant regional scale datasets to describe the strategic and operational contexts that each range manager must operate within. These sources of data and information can inform Defence land managers and enable land managers to reliably use robust information of changes and trends in vegetation condition (function, structure and composition) to achieve and maintain the objectives of each training area so that the relevant areas remain fit for purpose.

Under the Environment Protection and Biodiversity Conservation (EPBC) Act 1999, all Defence properties must be managed to prevent significant impacts to the environment, and consistent with managing land in accordance with recovery plans for rare and endangered species and communities and threat abatement plans aimed at protecting natural values. Defence properties also have an important complementary role (Margules and Pressey 2000) to the National Reserve System, which includes formally declared protected areas, indigenous protected areas and public-private covenanted reserves. For example, Shoalwater Bay Training Area in Queensland is the largest remaining area of sub-tropical coastal heathland on the Australian east coast—an ecosystem type which is relatively poorly protected in formal reserves (Keith et al. 2014). It is therefore not surprising that Zentelis and Lindenmayer (2015) and others suggest that, with appropriate integrated management, the Defence estate has potential to play critical complementary roles alongside the formal protected area estate (e.g. International Union for the Conservation of Nature (IUCN) protected areas categories I–IV).

5 Conclusions

Analyses using national vegetation datasets including the condition of native vegetation, landscape alteration levels and local comparisons between vegetation transformation within and outside training areas has shown the importance of the Defence properties in maintaining natural ecosystem services, particularly in more intensively managed and developed landscapes. We have shown that compared to other land uses the Defence estate is less modified and less fragmented. As a major land holding, the Defence estate also plays a valuable contribution in national vegetation accounting and this chapter highlights the need to regularly assess and report vegetation type, extent and condition across the Defence estate.

Acknowledgements We acknowledge the support given by Lucy Randall in providing access to the vegetation, land use and agro-climate datasets. Stuart Pearson commented on an earlier draft.

Disclaimer The views expressed are the authors' and not necessarily those of the Department of Defence. The Commonwealth of Australia will not be legally responsible in contract, tort or otherwise, for any statement made in this paper.

Appendix

Table 11.3 Relative area of defence facilities found in agro-climate regions compared with total area of all land uses

Class	Agro-climate region	Area of agro-climate class in Australia (ha)	Area of estate in agro-climatic class (ha)	Percent of total area of estate in agro-climate class compared to total area of all land uses	Number of facilities in agro-climatic class
1	Cold wet	3,890,325	-	0.00%	0
2	Temperate cool-season wet	28,980,115	95,767	0.33%	191
3	Mediterranean	63,468,160	73,356	0.12%	95
4	Temperate - sub-humid	14,145,313	6194	0.04%	13
5	Sub-tropical sub-humid	38,521,929	1669	0.00%	14
6	Sub-tropical moist	11,218,341	82,159	0.73%	119
7	Tropical warm season wet	82,476,523	1,783,448	2.16%	64
8	Tropical warm season moist	24,438,653	524,446	2.13%	117
9	Tropical wet	2,457,959	17,055	0.69%	16
10	Dry	98,835,343	907,841	0.18%	54
	Totals	768,432,663	3,491,934	0.45%	683

Source of agro-climate regions: (Thackway and Freudenberger 2016, Fig. 11.1)

Table 11.4 Relative area of major vegetation groups found in defence facilities compared with total area of all land uses

#	Major Vegetation Group (MVG) Name Native vegetation and naturally bare types except for groups 25, and 99	Area of MVG class in Australia(ha)	Area of MVG class in total estate(ha)	Percent of MVG class in estate compared to all land uses
1	Rainforests and vine thickets	3,646,905	29,158	0.80%
2	Eucalypt tall open forests	3,714,845	11,388	0.31%
3	Eucalypt open forests	23,425,402	155,548	0.66%
4	Eucalypt low open forests	1,167,389	7675	0.66%
5	Eucalypt woodlands	84,528,297	1,081,521	1.28%
6	Acacia forests and woodlands	34,156,656	18,659	0.05%
7	Callitris forests and woodlands	3,415,718	399	0.01%
8	Casuarina forests and woodlands	1,657,916	26,462	1.60%

(continued)

Table 11.4 (continued)

#	Major Vegetation Group (MVG) Name Native vegetation and naturally bare types except for groups 25, and 99	Area of MVG class in Australia(ha)	Area of MVG class in total estate(ha)	Percent of MVG class in estate compared to all land uses
9	Melaleuca forests and woodlands	8,142,192	197,651	2.43%
10	Other forests and woodlands	4,455,005	20,607	0.46%
11	Eucalypt open woodlands	46,422,956	260,311	0.56%
12	Tropical eucalypt woodlands/grasslands	13,572,816	215,152	1.59%
13	Acacia open woodlands	38,272,902	344,063	0.90%
14	Mallee woodlands and Shrublands	21,376,333	142,590	0.67%
15	Low closed forests and tall closed Shrublands	1,758,042	5072	0.29%
16	Acacia Shrublands	85,536,647	153,577	0.18%
17	Other Shrublands	12,268,546	18,109	0.15%
18	Heathlands	1,442,923	8882	0.62%
19	Tussock grasslands	52,714,741	51,528	0.10%
20	Hummock grasslands	137,284,140	375,554	0.27%
21	Other grasslands, Herblands, Sedgelands and Rushlands	4,962,995	8255	0.17%
22	Chenopod Shrublands, samphire Shrublands and Farblands	48,917,553	131,546	0.27%
23	Mangroves	1,049,111	15,197	1.45%
24	Inland aquatic - freshwater, salt lakes, lagoons	8,342,951	7335	0.09%
25	Cleared, non-native vegetation, buildings	103,263,048	97,752	0.09%
26	Unclassified native vegetation	179,277	124	0.07%
27	Naturally bare - sand, rock, claypan, mudflat	1,595,721	6689	0.42%
28	Sea and estuaries	89,509	6	0.01%
29	Regrowth, modified native vegetation	1,236,766	312	0.03%
30	Unclassified Forest	99,125	111	0.11%
31	Other open woodlands	16,960,581	99,181	0.58%
32	Mallee open woodlands and sparse Mallee Shrublands	2,077,874	675	0.03%
99	Unknown/no data	1,350,186	6588	0.49%
	Totals	769,085,068	3,497,677	0.45%

Source Major Vegetation Groups: Department of Environment and Energy (2016a and b)

Table 11.5 Relative area of vegetation condition classes found in defence facilities compared with total area of all land uses

Class	Vegetation condition classes (VAST)	Area of Vegetation condition class in Australia(ha)	Area of condition class in total estate(ha)	Percent of total area of estate in condition class compared to total area of all land uses
0	Naturally bare	11,782,000	10,166	0.09%
1	Residual or unmodified	336,914,000	2,158,121	0.64%
2	Modified	204,254,000	836,938	0.41%
3	Transformed	105,664,000	362,298	0.34%
4	Largely replaced and plant community is adventive	0	-	
5	Replaced and managed for intensive use production	108,666,000	96,355	0.09%
6	Replaced with man-made structures / infrastructure	953,000	8746	0.92%
	Totals	768,233,000	3,472,624	0.45%

Source vegetation condition: ABARES (2008)

Table 11.6 Relative area of landscape alteration levels found in defence facilities compared with total area of all land uses

Landscape Alteration Level classes	Area of LAL class in Australia(ha)	Area of LAL class in total estate(ha)	Percent of total area of estate in LAL class compared to total area of all land uses
Intact >90 retained	352,152,700	1,864,540	0.53%
Variegated >60–90 retained	21,096,700	1,515,306	0.47%
Fragmented 10–60 retained	56,073,600	79,001	0.14%
Relictual <10 retained	38,830,500	12,215	0.03%
Totals	68,153,500	3,471,062	0.45%

Source Landscape Alteration Levels: ABARES (2010)

References

ABARES. (2008). National Scale Vegetation Assets, States and Transitions (VAST Version 2) – 2008. Online at: http://www.agriculture.gov.au/abares/publications/display?url=http://143.188.17.20/anrdl/DAFFService/display.php?fid=pa_vast_g9abll0032008_11a.xml [Accessed 21 July 2017].

ABARES. (2010). A method for deriving maps of landscape alteration levels from vegetation condition datasets 2010. Online at: http://www.agriculture.gov.au/abares/publications/display?url=http://143.188.17.20/anrdl/DAFFService/display.php?fid=pe_brs90000004186.xml [Accessed 21 July 2017].

Bell, A. (1985). The conservation value of Defence training areas. *Ecos* 43 Autumn 17–23. Online at: http://www.ecosmagazine.com/?paper=EC43p17 [Accessed 21 July 2017].

Bowett, J., Davidson, A., & Danvers, T. (2012). Shoalwater Bay training area: Capability, conservation and collaboration. In P. Figgis, J. Fitzsimons, & J. Irving (Eds.), *Innovation for 21st century conservation* (pp. 142–147). Sydney: Australian Committee for IUCN. Online at: http://cmsdata.iucn.org/downloads/innovation_for_21st_century_conservation_low.pdf [Accessed 21 July 2017].

Bureau of Meteorology. (2013). *Guide to environmental accounting in Australia, Environmental Information Programme, Publication Series* (Vol. 3, p. 122). Canberra: Bureau of Meteorology.

COAG Standing Council on Environment and Water. (2012). *Australia's native vegetation framework*. Canberra: Council of Australian Governments, Department of Sustainability, Water, Population and Communities.

Department of Defence. (2009). *Defending Australia in the Asia Pacific Century: Force 2030*. Defence white paper 2009, Department of Defence, Canberra, Australia. p 122. Online at: http://www.defence.gov.au/whitepaper/2009/docs/defence_white_paper_2009.pdf [Accessed 21 July 2017].

Department of Defence. (2016). *Environment Strategy 2016*. Department of Defence, Canberra, Australia. Online at: http://www.defence.gov.au/estatemanagement/governance/Policy/Environment/Policy/EnvironmentStrategy2016.PDF [Accessed 21 July 2017].

Department of Environment and Energy. (2016a). *Australia - Pre-1750 Major Vegetation Groups - NVIS Version 4.2 (Albers 100m analysis product)*. Department of Environment and Energy, Canberra, Australia. Online at: http://www.environment.gov.au/fed/catalog/search/resource/details.page?uuid=%7B3B0B4F84-AA9E-40FB-B5C1-0DB9019CB1A1%7D [Accessed 21 July 2017].

Department of Environment and Energy. (2016b). *Australia - Present Major Vegetation Groups - NVIS Version 4.2 (Albers 100m analysis product)*. Department of Environment and Energy, Canberra, Australia. Online at: http://www.environment.gov.au/fed/catalog/search/resource/details.page?uuid=%7BA7C505C1-E11A-4D1A-8F9D-06AFB5E004A1%7D [Accessed 21 July 2017].

Fitzsimons, J., Pulsford, I., & Wescott, G. (2013). Linking Australia's landscapes: An introduction. In J. Fitzsimons, I. Pulsford, & G. Wescott (Eds.), *Linking Australia's landscapes lessons and opportunities from large scale conservation networks*. Collingwood, Victoria: CSIRO Publishing.

Hobbs, R. J., & McIntyre, S. (2005). Categorizing Australian landscapes as an aid to assessing the generality of landscape management guidelines. *Global Ecology and Biogeography, 14*, 1–15.

Keith, D., Lindenmayer, D., Lowe, A., et al. (2014). Heathlands. In D. Lindenmayer, E. Burns, N. Thurgate, & A. Lowe (Eds.), *Biodiversity and environmental change*. Melbourne: CSIRO Publishing.

Lesslie, R., Thackway, R., & Smith, J. (2010). *A national-level vegetation assets, states and transitions (VAST) dataset for Australia (version 2)*. Canberra: Bureau of Rural Sciences. Online at: http://data.daff.gov.au/data/warehouse/pe_brs90000004193/VASTv2Data_20100320_ap14.pdf [Accessed 21 July 2017].

Margules, C. R., & Pressey, R. L. (2000). Systematic conservation planning. *Nature, 405*(6783), 243–253.

McGarigal, K., Cushman, S. A., Neel, M. C., & Ene, E. (2002). *FRAGSTATS: Spatial pattern analysis program for categorical maps*. Amherst, MA: University of Massachusetts.

McIntyre, S., & Hobbs, R. (1999). A framework for conceptualising human effects on landscapes and its relevance to management and research models. *Conservation Biology, 13*, 1282–1292.

MPIG (Montreal Process Implementation Group for Australia and the National Forest Inventory Steering Committee). (2013). State of the Forest Report. Department of Agriculture and Water Resources, Canberra, Australia.

Mutendeudzi, M., Stafford-Bell, R. (2011). *Scientific information for making decisions about natural resource management – a report on the value, status and availability of key ABARES datasets*. ABARES technical report 11.2, Department of Agriculture and Water Resources Canberra, Australia.

Mutendeudzi, M., & Thackway, R. (2010). *A method for deriving maps of landscape alteration levels from vegetation condition state datasets*. Canberra: Bureau of Rural Sciences. Online at: http://143.188.17.20/data/warehouse/pe_brs90000004186/lalMethTechRpt_20091208_r1_ap14.pdf [Accessed 21 July 2017].

Natural Resource Management Ministerial Council. (2010). *Australia's Biodiversity Conservation Strategy 2010–2030*. Department of Sustainability, Water, Population and Communities, Canberra, Australia.

PGM Environment and Eco Logical Australia. (2014). Strategic Environmental Assessment of Defence Activities in the Great Barrier Reef World Heritage Area, 2014: Update Report. Prepared for the Department of Defence. pp 43. Online at: http://www.gbrmpa.gov.au/__data/assets/pdf_file/0008/248633/2014_Strategic-Environmental-Assessment_Defence.pdf [Accessed 21 July 2017].

Sbrocchi, C., Davis, R., Grundy, M., Harding, R., Hillman, T., Mount, R., Possingham, H., Saunders, D., Smith, T., Thackway, R., Thom, B., & Cosier, P. (2015). *Technical analysis of the Australian regional environmental accounts trial*. Sydney: Wentworth Group of Concerned Scientists. isbn:978-0-9944577-1-4.

State of the Environment report. (2011). Chapter 2.3 Vegetation. Independent report to the Australian Government Minister for Sustainability, Environment, Water, Population and Communities, Canberra, Australia.

Thackway, R., & Freudenberger, D. (2016). Accounting for the drivers that degrade and restore landscape functions in Australia. *Land, 5*(4), 40; 1–20, s1–s15. doi:https://doi.org/10.3390/land5040040.

Thackway, R., & Lesslie, R. (2006). Reporting vegetation condition using the Vegetation Assets, States and Transitions (VAST) framework. *Ecological Management & Restoration, 7*, S53–S62.

Thackway, R., & Lesslie, R. (2008). Describing and mapping human-induced vegetation change in the Australian landscape. *Environmental Management, 42*, 572–590.

Thackway, R., & Pearson, S. (2017). *Assessing changes in the extent and condition of native vegetation on the RAAF Base Amberley – a scoping study. Report prepared for the environment directorate* (p. 52). Canberra: Department of Defence. UNSW Canberra at the Australian Defence Force Academy.

Thackway, R., & Specht, A. (2015). Synthesising the effects of land use on natural and managed landscapes. *Science of the Total Environment, 526*, 136–152.

Yapp, G. A., & Thackway, R. (2015). Responding to change — Criteria and indicators for managing the transformation of vegetated landscapes to maintain or restore ecosystem diversity. In J. A. Blanco (Ed.), *Biodiversity in ecosystems - linking structure and function*. ISBN: 978-953-51-2028-5, InTech, Online at: http://www.intechopen.com/books/biodiversity-in-ecosystems-linking-structure-and-function/responding-to-change-criteria-and-indicators-for-managing-the-transformation-of-vegetated-landscapes [Accessed 21 July 2017].

Yapp, G., Walker, J., & Thackway, R. (2010). Linking vegetation type and condition to ecosystem goods and services. *Ecological Complexity, 7*(3), 292–301. https://doi.org/10.1016/j.ecocom.2010.04.008.

Zentelis, R., & Lindenmayer, D. (2015). Bombing for biodiversity—Enhancing conservation values of military training areas. *Conservation Letters, 8*(4), 299–305.

Chapter 12
Australia's First and Most Important War

Henry Reynolds

Geography has always featured in Australian historical writing. This was particularly so in the late nineteenth century and early twentieth century. Exploration was at the centre of popular history and featured in school text books. The voyages of European seamen were traced on maps of the continent. They were followed by the expeditions of the inland explorers. Atlases used in schools until the 1950's contained a series of maps depicting the progress of the settler's advance into Aboriginal Australia. On these maps black was succeeded by triumphant yellow until the last unknown regions were traversed by the intrepid frontiersmen. The explorers were succeeded in popular histories by the pioneers who were confronted by the land itself. They endured flood, drought and fire and in doing so developed characteristics which came to be seen as distinctly Australian. In popular literature, it was the Outback—not the coastal cities—which shaped the national ethos. Both sides of politics found their rural heroes. The (conservative) *right* celebrated the squatter. The (progressive or liberal) *left* venerated the semi-nomadic tribe of bush workers, shearers and their precocious unionism.

Distance was a constant theme (Blainey 1966), firstly in maritime and then inland narratives. One of the earliest celebratory stories was the crossing of the Blue Mountains in 1813. The largely fruitless search for navigable waterways was another common theme. The bullock dray, the Cobb and Co. coaches and the Afghan camel trains were embodied in popular history and literature. Australia's Federation itself was often seen as a victory over distance as much as an elaborate constitutional arrangement. But Australia's position in the world mattered as well. Distance was a double-headed problem. There was the vast distance from the European homelands which only the well-heeled could ever hope to revisit. And then there

H. Reynolds (✉)
Aboriginal Studies, Global Cultures & Languages, University of Tasmania, Sandy Bay Campus, Hobart, TAS, Australia
e-mail: Henry.Reynolds@utas.edu.au

S. Pearson et al. (eds.), *Australian Contributions to Strategic and Military Geography*, Advances in Military Geosciences,
https://doi.org/10.1007/978-3-319-73408-8_12

was the even more challenging proximity of East and South-East Asia with their vast but little-known populations. The threat was compounded by the overwhelming fact that apart from coastal Queensland, the whole tropical north was home to few resident white Australians and even in the small towns they were outnumbered by Aborigines and assorted Asian settlers (Reynolds 2003b).

But while geography has been widely employed to explain the history of settlers in Australia it has rarely been used in a similar way to interpret indigenous history in general and frontier warfare in particular. But both the size and diversity of the continent need to be closely considered. The relevant facts are so obvious they are often overlooked. They need spelling out. Aborigines occupied every part of the continent even though population densities varied in response to the local availability of food and water. They were far more widely distributed than the Europeans in the nineteenth century or even today (Tindale 1974). They utilised a much wider range of indigenous animals and plants than typical settlers and their knowledgeable use of fire effects is remarkable (Gammage 2011). As a result, Australia's indigenous peoples have survived and flourished in country that Europeans have never been able to exploit.

The pattern of white settlement was determined as much by geography as by cultural preferences. In the age of sail, European ships had to travel far to the south around one of the two great Capes to enter the Indian or Pacific Oceans. The voyage east around the Cape of Good Hope was the far more popular option and from there the westerly winds facilitated an easy, if often boisterous, passage to southern Australia. So, settlement began and was concentrated in the southern and south-eastern fringes of the continent, where European crops flourished and where familiar practices of husbandry and cultivation could be utilised.

One unintended consequence of this pattern of discovery and settlement was that the British had their earliest contact with the Aboriginal nations in the south which had no experience of contact with foreign visitors. The Torres Strait Islanders or especially the Yolngu of Arnhem Land in the north, for example, had many years of intimate relations with the Indonesians before they were forced to come to terms with the Europeans (MacKnight 1976). By contrast the Tasmanians living on the Island's south–east coast, were visited by ten British and French expeditions between 1772 and 1803, but had lived in complete isolation from the rest of humanity for at least 300 generations. The appearance of alien foreigners was far more astonishing, traumatic and arguably disabling for the tribes living around the continent's south coast than it was for their contemporaries in the north, who had also developed greater immunity to the bacteria and viruses carried in from outside.

But, even in the south, Aboriginal experience of Europeans varied. There was a great contrast between those nations that lived on or near the sites chosen for the original colonial or penal towns like Sydney, Hobart, Perth and Adelaide, and numerous other ancillary settlements. In these cases, the Aboriginal communities were immediately overwhelmed by what, in their experience, were huge crowds of people along with their domestic animals and innumerable possessions. On the other hand, Aborigines in the vast hinterlands had warning of the arrival of the strangers, sometimes many years before they actually met them. Information about them was passed back beyond the frontiers of settlement. and parties almost certainly travelled in towards the white men to examine them for themselves. Meanwhile feral European animals (cattle, horses, dogs and cats) strayed out into the interior well in advance of their erstwhile owners. Pieces of iron and glass, knives and small axes were traded back into the interior to be incorporated into traditional tool kits (Reynolds 2006).

The single most important characteristic of European settlement, is that it took so long to complete the occupation of the whole Australian continent. So much of it was completely inhospitable (like the vast central deserts), difficult to access or to turn to economic use—places as various as rugged highlands or wetlands and mangrove tangled foreshores. Consequently, those traumatic moments of first contact which began in the eighteenth century continued until the middle of the twentieth century. The last people leading a traditional desert life were not contacted until the 1960's (Davenport et al. 2005). The process of contact and concomitant conflict accompanied the full stretch of white Australian history. Modern urban Australia lives side by side with Aboriginal communities which still speak their own languages and practice many traditional customs. They occupy large areas of the north and centre, and regionally outnumber resident Europeans; few of whom have ever lived there. And these are often frontier areas of great strategic significance, like the Torres Strait and the Tiwi Islands.

Another consequence of the long, drawn-out process of occupation and settlement was that frontier violence continued for 140 years, making it one of the most enduring features of national history. It began with the settlement of the Hawkesbury River Valley in the 1790s and persisted until the massacre of a large party of Aborigines at Coniston Station in Central Australia by a police party in 1928. Very few white men were ever brought to trial for killing Aborigines. The consequences of this long history of unpunished violence has not been fully considered by Australians. It certainly influenced general attitudes to the Aborigines, and nourished the racism which inspired the communal dedication to the White Australia Policy—which persisted until the 1970s. But the impact on Aboriginal society was certainly greater. There are old people still alive who, as children, heard eye witness accounts of brutal killing and certainly knew of children taken away by police—never to return. Until quite recently, Aboriginal children were taught to never 'back chat' a white man or even look too directly at one. 'Cheeky Black Fellas', it was explained, were likely to be bashed or even killed. Deference and down-cast eyes were prudent strategies to avoid trouble. Consequently, residual hatred of white people sometimes survives like a fire deep underground (Reynolds 1987).

The longevity of frontier warfare ensured that it varied from place to place and over time. By the late nineteenth century colonial writers could look back over almost a century of violence. The pioneer ethnographers Lorimer Fison and A.W. Howitt observed in 1880 that:

> It may be stated broadly, that the advance of settlement has, upon the frontier at least, been marked by a line of blood. The actual conflict of the two races has varied in intensity and in duration, as the various tribes have themselves in mental and physical character. … But the

> tide of settlement has advanced along an ever-widening line, breaking the tribes with its first waves and overwhelming their wrecks with its flood.(Howitt and Fison 1889)

Six years later a second ethnographer, Edward Curr (1886), wrote another overview of frontier warfare which was based on a wide correspondence with informants all over Australia:

> In the first place the meeting of the aboriginal tribes of Australia and the White pioneer, results as a rule in war, which lasts from six months to ten years, according to the nature of the country, the amount of settlement which takes place in a neighbourhood, and the proclivities of the individuals concerned. When several squatters settle in proximity and the country they occupy is easy of access and without fastnesses to which the blacks can retreat, the period of warfare is usually short and the bloodshed not excessive. On the other hand, in districts which are not easily traversed on horseback, in which the Whites are few in number and food is procurable by the Blacks in fastnesses, the term is usually prolonged and the slaughter more considerable. (Curr 1886)

The Black War in Tasmania between 1826 and 1831 illustrated many of Curr's observations. More European lives were lost there than anywhere else in Australia. About 250 settlers were killed in Aboriginal attacks. The seriousness of the war was indicated by the response of the administration of Governor George Arthur, which conducted the so-called Black Line in 1828 with the intention of driving the most hostile bands out of the settled districts and onto the peninsulas in the Island's south-east. The record on a Surveyor General's map (Fig. 12.1) shows the arrangements of the force. The whole community was organised. Five hundred and fifty British soldiers took part along with 1700 armed ancillaries. The whole host was out in the field for 6 weeks, most of them removed from their workplaces. The rugged, unmapped country and inclement winter weather militated against the venture, which was almost a complete failure. It cost as much as a year's government expenditure.

But the Governor clearly saw Aboriginal hostility as a serious threat to the colony. Most of the military officers, had either personally experienced guerrilla warfare while campaigning in Spain (Peninsular War 1807–1814) or, like Arthur, they knew about it. What they most feared, quite reasonably, was that the Aborigines would undertake a systematic campaign of burning the impending wheat crop on which the colony depended. And the rural workforce itself was so vulnerable. As assigned convict servants, they were compelled to work in vulnerable locations which free labourers would have avoided. They were not normally trusted with guns and were almost always on foot. Tasmania had not experienced the rapid democratisation of horse riding that was unfolding on mainland Australia. And there was little the government could do. It was so hard to hit back. The Tasmanians had no villages to storm, no forts to capture, no domestic animals to sequester, no crops to burn or wells to poison.

In Curr's terminology there were "fastnesses" everywhere. The settlers occupied the valleys and narrow plains. However, interlocking, forested hills and attendant mountains defined every horizon. The Aborigines knew their country, the Europeans did not and there were no detailed maps. The hills provided ideal locations for the Aborigines to watch over the settler's farms, either to launch attacks or escape from

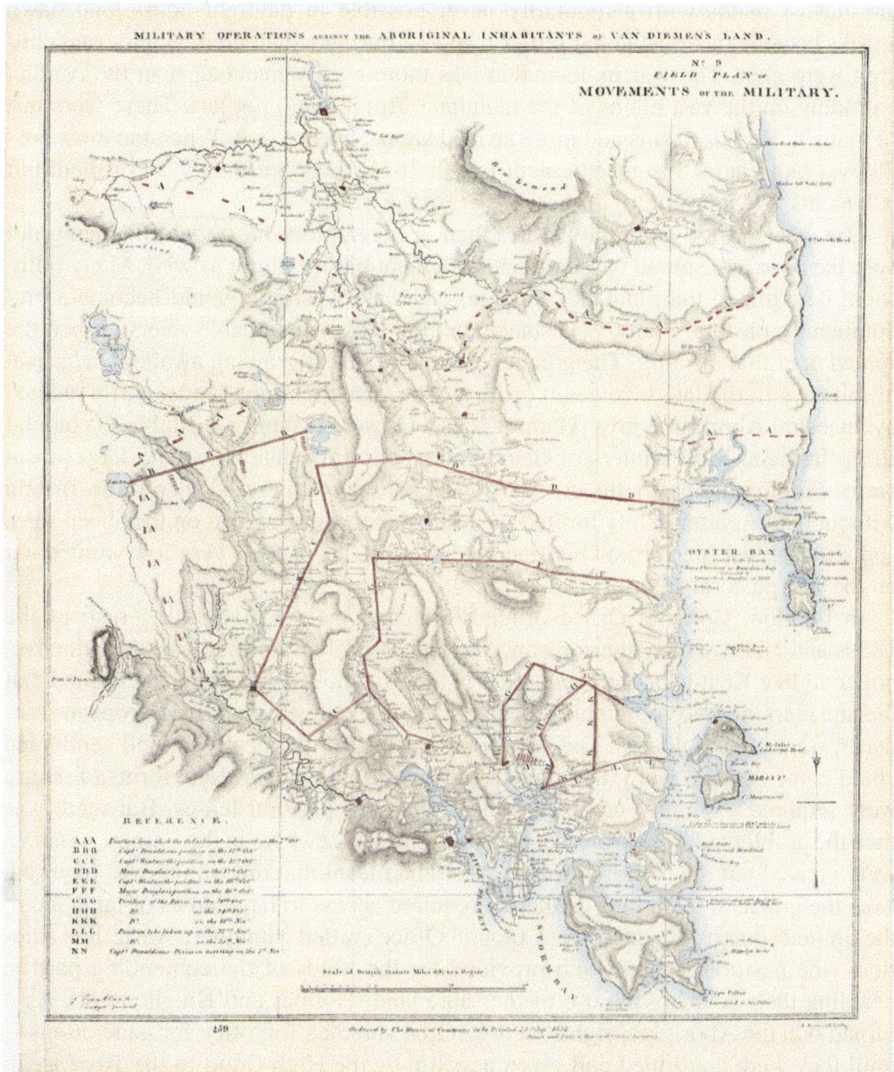

Fig. 12.1 Military Operations against the Aboriginal Inhabitants of van Diemans Land No 9 Field Plan of Movement of the Military) recording front lines in October and November 1831 (Source G Frankland Surveyor General National Library of Australia http://nla.gov.au/nla.obj-274524964)

pursuing parties which had to advance in full view along the valley floors. Both water and game were available in the hills, although the war bands increasingly adopted flour and potatoes stolen from European huts as hunting itself became a hazardous occupation requiring access to potentially dangerous open country.

But where the Aborigines were vulnerable and where they were most often attacked was around their camp fires at night. Fire enabled parties of settlers to find the elusive bands with a regularity never possible in daylight hours, and dawn attacks became the most lethal settler response during which men, women and children were shot. The war in Tasmania was more evenly matched than the conflict unfolding on the vast plains of the mainland. But it could not last. There were just so many more Europeans and more arrived almost every week. When the most hostile war band agreed to a settlement late in 1831, there were only 26 of them left (Clements 2013; Ryan 2012).

The war and its geography were, then, a peremptory master—influencing not only the pace and spread of European settlement but its nature as well. Many influences determined that, unlike Tasmania, mainland Australia would become a predominantly pastoral country. Distance and the lack of navigable rivers curbed the spread of arable farming. The great expansion of wheat growing awaited the arrival of railways in the late nineteenth century. The vast open grasslands, partly created by ancestral Aboriginal fires (Gammage 2011), were instantly available to colonial sheep farmers. Mild winters obviated the expense of winter feed and large warm barns. Wool could cover the cost of transport to port cities and then on to Britain without deterioration. Only limited unskilled labour was needed on the sheep runs, and shearing was performed by itinerant shearers. There were very few women out on the vast pastoral frontier.

In London, Colonial Office officials watched the great squatting rush of the 1830s and 1840s with gathering concern. The Australian frontier was so distinctive; not at all like Kentucky, they remarked in their memos. Geography determined that the squatters needed to be able to graze their flocks over what, by European standards, were huge holdings. The only comparisons the classically trained gentlemen could come up with were the nomadic tribes of Central Asia. New forms of tenure were required—initially grazing licences and then pastoral leases. But there was then the problem of how to protect the Aborigines. Elsewhere in the Empire, reserves were created but the nature of inland Australia meant that innovation was required. Like the graziers, the hunter-gatherers required access to large areas of land and to the limited sources of water. The Colonial Office crafted a uniquely Australian solution—the pastoral lease, which provided for the needs of the contending parties, enabling them to share the use of the same land (Holmes and Knight 1994). As it turned out, the Aboriginal rights were ignored and then forgotten for generations—until they were confirmed and given new life by the High Court in the *Wik* case in 1996 (Reynolds 2003a).

The well-intended reform in the Colonial Office had little effect on the conflict, which followed the white frontiersmen into every new district occupied from the 1830s to the 1920s. They led their sheep and cattle from northern New South Wales in the 1840s up onto the Darling Downs and then into Central Queensland in the 1860s. By the 1880s Queensland cattle men had pushed up into Cape York, and then across the tropical savannah to meet in the Kimberly with the West Australian pioneers who had brought their sheep flocks north from the Gascoyne and the Pilbara.

Conflict with resident Aborigines was ubiquitous and inescapable. The flocks and herds crowded onto the land long cleared by fire for efficient hunting. Above all, they monopolised the usually limited water sources. In the long dry season of the tropical north, conflict over water was literally a matter of life and death. In most districts the Aborigines outnumbered the small pastoral workforce. But by the middle of the nineteenth century, the balance of power had shifted decisively towards the invaders. They lived on horseback. Their rifles and revolvers made them invulnerable, and they had inherited bush skills and an ethic of brutal assertion. They were without women and took Aboriginal girls by force or in steeply one-sided negotiation.

But once resistance had been crushed the large scale killing had to stop. The Aborigines were needed to provide the labour on stations so far away from urban or even small-town Australia where white workers were scarce, unreliable and expensive. By 1900 as many as 10,000 Aboriginal men and women worked in the pastoral industry in northern Australia. They brought valuable skills with them. They knew their country far better than the white boss. Their tracking capacity was irreplaceable on the vast unfenced stations. And the young people became legendary horse men and women. But the work of the stations was dictated by the climate. During the dry season the Aborigines lived in station camps, a discrete distance from the white homesteads. The young men and women worked for the white boss, while often large numbers of dependants lived in the camp, periodically given a slaughtered animal. But once the wet arrived all work ceased and the whole Aboriginal camp returned to the bush to resume their traditional life; the calendar of ritual events modified to fit in with the demands of station work. This pattern of station life persisted in much of northern Australia until the 1960's (May 1994; Roberts 2005).

Frontier conflict was a feature of Australian life for more than a century. There was a considerable death toll. Several thousand settlers died and many more Aborigines, although we will never know how many. The most recent assessment is that at least 30,000 were killed in skirmishes with the settlers and various official para-military forces—in particular the Queensland Native Police, which patrolled the northern frontier for fifty years (Bottoms 2013; Foster and Attwood 2003; Nettlebeck and Foster 2007; Nettlebeck and Foster 2012; Owen 2016; Richards 2008). Despite the long duration of frontier warfare and the great diversity of country, there was a high degree of continuity in tactics and operation of the conflict. Land was at the centre of contention. Wide personal variation did not change the fundamental dynamic. The Europeans were willing to endure the hardship and danger of frontier life to assert their control over the land and its varied resources. And at the heart of their endeavour was the belief that the Aborigines did not own their homelands. Australia was legally a *terra nullius* and that definition was embedded in colonial law from the first days of settlement—until 1992 when it was overturned by the High Court of Australia in the *Mabo* case. The Aborigines had never accepted the law that had in theory dispossessed them. Coming to terms with the white men did not include capitulation. Wherever possible, they remained on their homelands or lived close by educating their children about their birthright. In the more remote parts of Australia, few Europeans ever appeared or didn't stay long when they did. Over much of the continent, geography stood in the way of settlement. The

Aborigines were able to re-assert their ownership once the land rights movement arose in the 1960s and 1970s.

When native title was eventually incorporated in Australian law in 1993, communities all over Australia prepared claims which went before the Native Title Tribunal or, in numerous cases, ended up in the courts. Up to the present, more than 350 claims have been accepted covering 40% of the continental land mass. Title has been returned to traditional owners. The 200-year saga of assertive European colonisation, which had faltered in the face of forbidding geographical challenges, had finally been brought to an end. Because so much of the continent had been unsuitable for European settlement and economic exploitation large areas remained as unallocated Crown land and consequently available for claim by traditional owners. And even on the vast areas of pastoral savannah residual Aboriginal rights were affirmed by the High Court in the *Wik* case of 1996.

The homelands and outstations movement ran parallel with these developments. From the 1980s, Aborigines in remote areas began to move away from missions and settlements back to their traditional homelands. The Mabo judgement hastened the exodus. In the Northern Territory there are currently 73 established remote communities with as many as 10,000 residents. There are many more small outstations in both the Territory and the north of Western Australia. This has dramatically changed the demography of the whole of northern Australia. It also has important, if little appreciated, geo-political implications.

One of the most enduring strategic concerns that Australians have contended with arises from the perceived problems created by the 'empty' north (See for instance McGregor 2016). For a long time, it was associated with the belief that the resident Aboriginal population would slowly decline and eventually disappear. These views were still commonly held until the middle of the twentieth century. The Aboriginal resurgence has meant that the population in remote areas has recovered and there has been a re-occupation of lands that had often been abandoned for a generation or more. This has enormous advantages in relation to the proper management of country. But it also signals to the wider world that Australia is in effective occupation of even the most remote parts of the continent. And that is important. Indigenous land owners have what might be termed a double legitimacy. They have both the legal standing which comes from Australian citizenship as well as the additional prestige of traditional ownership which has quite distinctive recognition enshrined in international law (ILO 1989; UN 2007).

Any serious consideration of frontier conflict necessarily provokes the question of how it should be incorporated in the influential national narrative about Australia's serial engagement in overseas wars. Several obstacles have historically stood in the way of mature reflection. For much of the first part of the twentieth century the Aborigines were written out of national history and little attention was given to conflict along the ragged fringes of settlement. When, in the 1970's, revisionist historians began to draw attention to the long history of skirmishing it was common to interpret the conflict as arising from theft or trespass; to disputes about women or to personal revenge. It lacked the political or strategic dimension which would lift it above the level of petty crime, or even banditry or piracy. The larger questions of property and sovereignty were not thought to be in contention because both questions had been decided at the beginnings of settlement.

As with so many other matters the *Mabo* judgement of 1992 impelled a reorientation of historical interpretation. With the rejection of *terra nullius*, the Court recognised that in 1788 the Aboriginal nations were both in possession of the land and the legitimate owners in accordance with their own laws and customs. Although the contentious question of sovereignty was not considered, the implication was that the land-owning nations also, of necessity, must have exercised a form of it over their own contained territories. The implication for the writing of history was inescapable. Frontier conflict was inescapably about the ownership and control of land right across one of the world's great land masses. The magnitude of the matters under contention meant that frontier conflict had to be radically reconsidered. It had to be elevated into the realm of warfare. It was profoundly political. And, that being the case, it had to be considered as a war of real consequence regardless of how it was fought. And, more to the point, because it was conflict about Australia and fought all over the continent it was arguably Australia's most important war.

References

Blainey, G. (1966). *The tyranny of distance: How distance shaped Australia's history*. Melbourne: Macmillan.

Bottoms, T. (2013). *The conspiracy of silence: Queensland's frontier killing times*. Sydney: Allen & Unwin.

Clements, N. P. (2013). *Black war*. Brisbane: University of Queensland Press.

Curr, E. M. (1886). The Australian race: Its origin, languages, customs, place of landing in Australia, and the routes by Which It Spread Itself (Vol. 1). Melbourne: Government Printers.

Davenport, S., Nixon, Y., & Johnson, P. (2005). *Cleared out: First contact in the Western Desert*. Canberra: Aboriginal Studies Press.

Foster, S. G., & Attwood, B. (Eds.). (2003). *Frontier conflict: The Australian experience*. Sydney: National Museum of Australia Press.

Gammage, W. (2011). *The biggest estate on earth: How Aborigines made Australia*. Sydney: Allen & Unwin.

Holmes, J. H., & Knight, L. (1994). Pastoral lease tenure in Australia: Historical relic or useful contemporary tool? *The Rangeland Journal, 16*, 106–121.

Fison, L., & Howitt, A. W. (1880). Kamilaroi and Kurnai, Facsimile edition, Oosthout,1967.

ILO. (1989). *Convention concerning indigenous and tribal peoples in independent countries; Part 2*. Geneva: International Labor Organisation.

MacKnight, C. (1976). *The voyage to Marege: Macassan trepangers in northern Australia*. Melbourne: Melbourne University Press.

May, D. (1994). *Aboriginal labour and the cattle industry: Queensland from white settlement to the present* (Vol. 13). Brisbane: CUP Archive.

McGregor, R. (2016). *Environment, race, and nationhood in Australia: Revisiting the empty north*. New York: Palgrave Macmillan.

Nettlebeck, A., & Foster, R. (2007). *In the name of the law*. Adelaide: Wakefield Press.

Nettlebeck, A., & Foster, R. (2012). *Out of the silence: The history and memory of South Australia's frontier wars*. Adelaide: Wakefield Press.

Owen, C. (2016). *"Every Mother's son is guilty": Policing the Kimberley frontier of Western Australia, 1882–1905*. Perth: University of Western Australia Press.
Reynolds, H. (1987). *Frontier: Aborigines, settlers and land*. Sydney: Allen and Unwin.
Reynolds, H. (2003a). *The law of the land*. Ringwood: Penguin.
Reynolds, H. (2003b). *North of Capricorn: The untold story of Australia's north*. Sydney: Allen and Unwin.
Reynolds, H. (2006). *The other side of the frontier: Aboriginal resistance to the European invasion of Australia*. Sydney: UNSW Press.
Richards, J. (2008). *The secret war: A true history of Queensland's native police*. Brisbane: University of Queensland Press.
Roberts, T. (2005). *Frontier justice: A history of the Gulf country to 1900*. Brisbane: University of Queensland Press.
Ryan, L. (2012). *Tasmanian Aborigines: A history since 1803*. Sydney: Allen & Unwin.
Tindale, N. B. (1974). *Aboriginal tribes of Australia: Their terrain, environmental controls, distribution, limits, and proper names*. San Francisco: University of California Press.
UN UNGA. (2007). *United Nations declaration on the rights of indigenous peoples*. New York: UN General Assembly.

Chapter 13
The Climate-Security Teleconnections of El Niño Southern Oscillation

Michael Thomas

1 Introduction

This chapter examines the security implications of El Niño Southern Oscillation (ENSO) events in the context of climate change. The first part briefly describes the physical basis of ENSO, and conducts a select review of literature suggesting climate change will serve as a catalyst for more intense and more frequent extreme El Niño and La Niña events. In developing this foundation, the latter part of this chapter examines the security implications of extreme ENSO events. Three such events are analysed—the 1982/83 El Niño, the 1997/98 El Niño and the 2015/16 El Niño—and the major human and national security consequences experienced across the planet are identified.

The case studies examined here explore the concept of security teleconnections that arise from ENSO events; whereby an extreme climatic phenomena or anomaly in one geographical location can be a cause-for, or catalyst-of, the unleashing of societal disruption, instability, violence or conflict in a distant (seemingly unrelated) part of the planet. In this regard, extreme climate teleconnection events can be a trigger for simultaneous instances of insecurity around the globe, profoundly impacting all levels of security (human, national, regional and, in extremis, global security).

Although teleconnections are well established concepts in meteorology and the earth sciences, they remain under-developed and largely absent concepts in security studies literature. This chapter argues that understanding climate-security teleconnections is important for contemporary analysts and policy-makers in seeking to better understand, interpret and even anticipate the geo-strategic security landscape

M. Thomas (✉)
Non-Resident Fellow, The Center for Climate & Security, Washington DC, USA
e-mail: mthomas@climateandsecurity.org

S. Pearson et al. (eds.), *Australian Contributions to Strategic and Military Geography*, Advances in Military Geosciences,
https://doi.org/10.1007/978-3-319-73408-8_13

of the Anthropocene. The chapter concludes by describing how future ENSO events, supercharged by climate change, may trigger widespread regional security consequences.

2 El Niño Southern Oscillation (ENSO) and Teleconnections

ENSO refers to periodic departures from expected sea surface temperatures (SST) across the equatorial Pacific and of atmospheric pressure differences between Darwin (Australia) and Tahiti (French Polynesia). ENSO events are characterised by three-to-seven year oscillations between neutral or 'normal' conditions, and between La Niña (cooler) and El Niño (warm) conditions. It is the dominant mode of global climate variability and the single most influential planetary climate pattern (Holbrook et al. 2009; Lindsey 2016; BoM 2017a). Fig. 13.1 illustrates these three states. Under normal conditions (1a), equatorial trade winds (via the Walker Cell) blow from east to west, causing warm water to gather in the western Pacific and for cold water to be upwelled along the eastern Pacific. During La Niña conditions (1b), equatorial trade winds of the Walker Cell are strengthened, resulting in a greater volume of warm water being pushed further west, toward the Maritime Continent, and an expansion of cold water across the eastern Pacific. Under El Niño conditions (1c), the Walker Cell weakens sufficiently such that the warm water now shifts eastwards towards the South American coastline, in the process curtailing the upwelling process prevalent in the other phases.

As a coupled atmosphere-ocean phenomenon, ENSO events are governed by interactions between these mediums. In the atmosphere—and as alluded above—ENSO is a measure of the difference in atmospheric pressure between Darwin and Tahiti, and is denoted by the Southern Oscillation Index (SOI) that provides an indication of the development and intensity of El Niño or La Niña events (BoM 2017a) (Box 13.1).

In the ocean, ENSO is monitored by sea surface temperatures in the regions, or 'boxes', located across the equatorial Pacific (Fig. 13.2). In this instance, El Niño (or La Niña) events are further characterised as being five consecutive 3-month running mean of SST anomalies in either the Niño 3, 3.4 or 4 region that are above (below) the threshold +0.5 °C (−0.5 °C). This standard of measure is known as the Oceanic Niño Index (ONI) (NOAA 2017b). Through studying SSTs in different regions across the Pacific, scientists have been able to discern different teleconnections and climate impacts arising from El Niño / La Niña events (Lee and McPhaden 2010).

Teleconnections can be defined as oceanic and atmospheric interactions that occur across widely separated regions; ensuring that the effects of El Niño and La Niña reverberate around the planet. In this manner, ENSO's fingerprints are to found on virtually every major planetary climate system including the Indian Monsoon (Ashok et al. 2004), the Gulf Stream (Taylor et al. 1998), Antarctic Circumpolar Wave (Peterson and White 1998) and Antarctic Dipole (Liu et al. 2002), Arctic

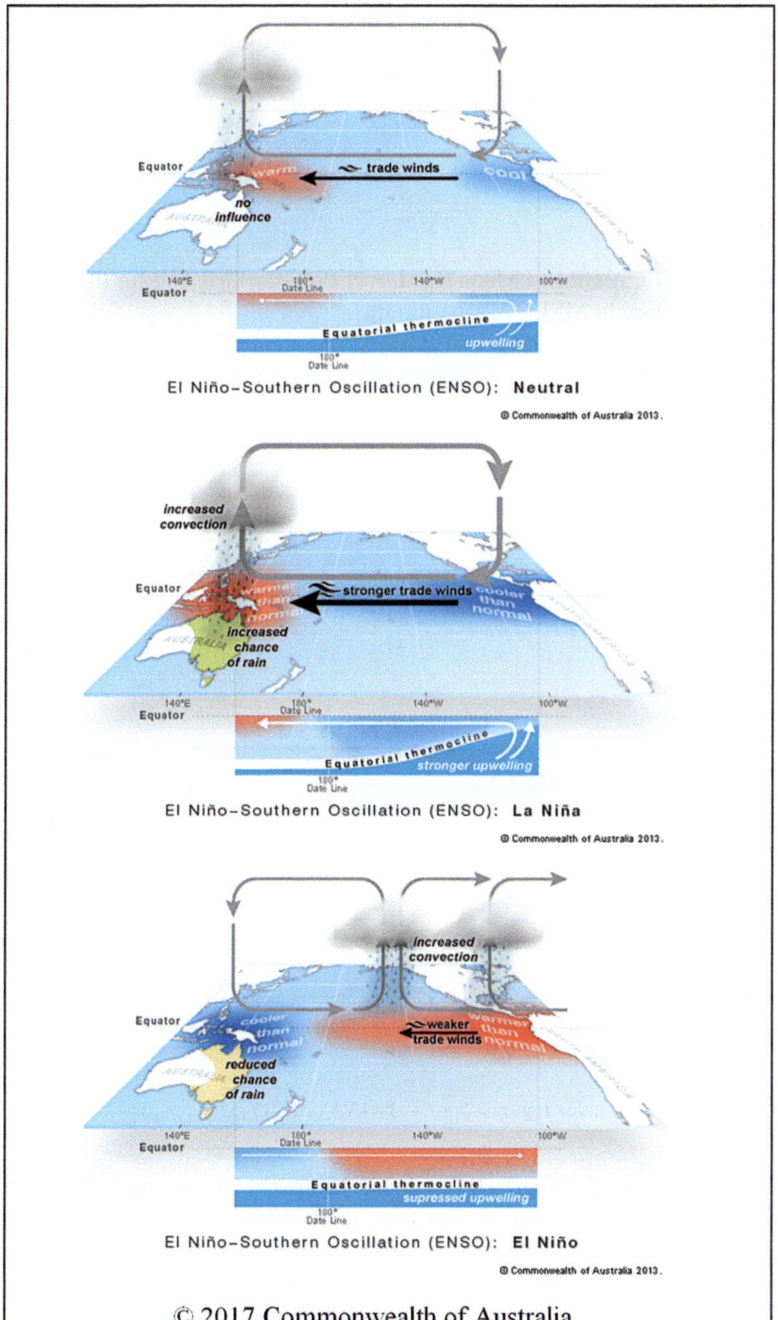

Fig. 13.1 ENSO cycles across the equatorial Pacific. Top-Bottom: 'Neutral', 'La Niña' and 'El Niño' (Source Commonwealth 2013)

Box 13.1 Australian Southern Oscillation Index Formula

There are various methods of calculating SOI. The Australian Bureau of Meteorology is the 'Troup SOI':

$$SOI = 10 \frac{[Pdiff - Pdiffav]}{SD(Pdiff)}$$

Where:

- Pdiff = (Average Tahiti Mean Sea Level Pressure (MLSP) for the month) - (average Darwin MSLP for the month),
- Pdiffav = long term average of Pdiff for the month in question, and
- SD(Pdiff) = long term standard deviation of Pdiff for the month in question.

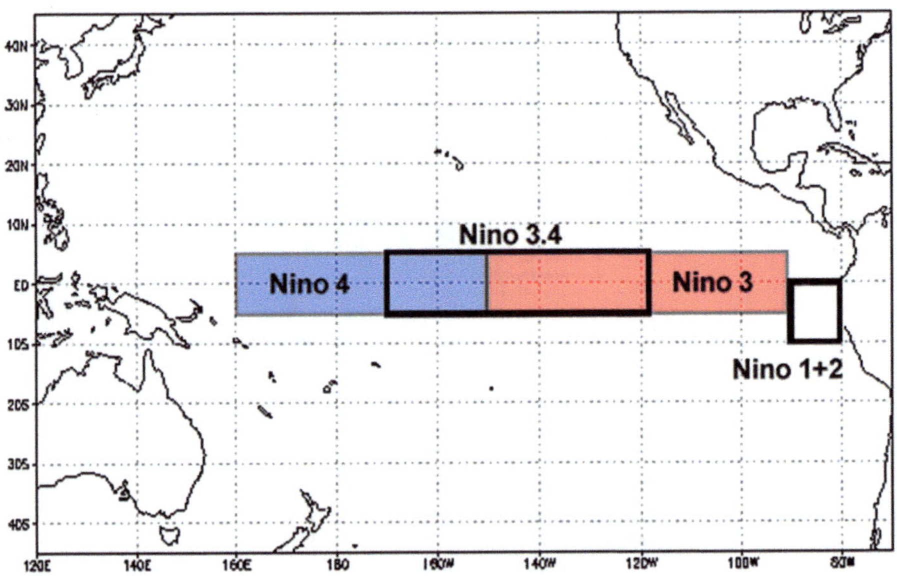

Fig. 13.2 Oceanic Niño index regions. *Source*: NOAA (2017c)

Oscillation (Lim and Schubert 2011), Asian Monsoon (Yuan and Yang 2012), and East African climate variability (Plisnier et al. 2000) to name a few. Through these teleconnections, ENSO events wield significant influence on regional climate and weather. During the months December through February, for example, El Niño brings warm and often dry (drought) conditions across Australia, South East Asia, much of South and far-north America, as well as Southern Africa and parts of the Sahel (WHO 2015). Simultaneously, it also brings wet conditions across much of southern continental United States and parts of East Africa. Under La Niña conditions for the same months, conditions are largely reversed (Lindsey 2016).

Not surprisingly, the ubiquity of El Niño and La Niña also warrants large influence on human societies and natural systems, affecting everything from the price of coffee in South America (Ubilava 2012) to finch productivity in northern Australia (Grant et al. 2000). Perhaps more ominously, and pertinently for this readership, El Niño and La Niña also have major food, water, economic and human security consequences through increased frequency and intensity of catastrophic floods, fires, drought, mass coral bleaching, heat waves, ecological and habitat destruction, and associated extreme weather events and large-scale natural disasters. To take one example from this list, some estimates place ENSO events to account for as much as 20% of global commodity price inflation, directly and significantly dampening world economic activity (Brunner 2002). When such phenomena recur and become the 'new normal', humanitarian crisis and regional instability might take hold as economies implode and governments struggle to provide basic services.

In this context, 'teleconnections' have meaning beyond climate and weather systems, and can also be conceived as having security consequences. Thus, climate-security teleconnections may arise when an extreme climatic phenomena or climate anomaly in one geographical location can be a cause-for, or catalyst-of, the unleashing of societal disruption, instability, violence or conflict in a distant part of the planet. To examine this phenomena in more detail, attention now turns to indications that climate change may be increasing the intensity and frequency of extreme El Niño and La Niña events, thereby worsening the security teleconnection consequences of such events.

3 Climate Change and "Extreme" ENSO Events

Since at least the 1990s, scientists have projected that climate change will engender more 'extreme' and even 'super' El Niño and La Niña events (Trenberth and Hoar 1997; Timmermann et al. 1999; Lee and McPhaden 2010; Cai et al. 2014; Cai et al. 2015). How are such events defined, what are their characteristic properties, and how do they differ from regular ENSO? To address these questions, several points merit discussion. Firstly, the term 'super' or 'extreme' applies (in this Chapter at least) to contemporary ENSO phenomena that operate over *inter-annual* cycles, as distinct from some definitions that apply the term to *millennial-scale* patterns (Stott et al. 2002). Secondly, because various methods exist by which ENSO may be measured, there is no consistent approach to defining the particular attributes or characteristics of such events. As a result, there is an inconsistency as to what events have actually been labelled as being 'extreme' or 'super' (see Box 13.2). Thirdly, there also appears no consolidated explanation in the literature on the quantitative difference between a 'super' versus 'extreme' event. Is a 'super' El Niño (or La Niña) in a category of its own, above 'extreme'? Or (and, as appears apparent) are the terms used interchangeably by different authors using different methods? Fourthly, while most definitions focus on the intensity of El Niño and La Niña events, it is conceivable that due consideration should also be given to the duration of an event.

Box 13.2 Defining Extreme ENSO Events

Extreme/Super El Niño. In her book Super El Niño, Li–Ciao Hong (2016) defines a 'super' El Niño as being of sufficiently 'large amplitude difference' above regular El Niño's or more precisely, of more than two standard deviations from the mean, and identifies 1972/73, 1982/83 and 1997/98 as candidate years. Conversely, the climate scientist Kevin Trenberth (2016) identifies three 'super' El Niño years as 1982/83, 1997/98 and the 2015/16 events (thereby discounting the 1972/73 event). Cai et al. (2014), likewise identify the 1982/83 and 1997/98 El Niño events, but rather refers to them as being 'extreme' events characterised as occurring when 'a massive reorganisation of atmospheric convection takes place such that [DJF] Niño3 rainfall exceeds 5 mm per day'. Institutions by contrast, such as the Australian Bureau of Meteorology (BoM) and National Oceanic and Atmospheric Administration (NOAA), generally avoid use of the terms 'extreme' or 'super' and rather classify ENSO events in terms as being 'Weak, 'Moderate', 'Strong' or 'Very Strong', according to set criteria. In more than 100 years of records, the Australian BoM and NOAA have identified three 'very strong' El Niño's (1982/83, 1997/98 and 2015/16).

Extreme/Super La Niña. The concept of a super or extreme El Niño gives rise to the question: what constitutes a super or extreme La Niña? Once again, whilst descriptions of 'extreme' La Niña events exist, there appears no widespread use of them being defined as 'super'. Cai et al. (2015), for example, define 1988/89 and 1998/99 as 'extreme' La Niña events and technically define them as being 1.75 standard deviation from mean SST in the Niño4 region. By contrast, BoM and NOAA have not identified any 'very strong' La Niña events.

In this regard, while the extended El Niño of the early-mid 2000s (which coincided with what was known in Australia as the Millennium Drought) did not attain the technical prerequisite for being 'extreme', its extended presence ensured a significant drying of the continent such that it severely impacted water, food and economic security (van Dijk et al. 2013). In this context, the magnitude of an ENSO event—a combination of *duration* and *intensity*—can also be important. Related to this is the unpredictability of ENSO events, such that a large (or 'extreme') El Niño or La Niña may not necessarily portend disaster. Indeed, on occasion, extreme ENSO may in-fact be uneventful or have an overall *positive* impact (Laosuthi and Selover 2007; Rosenzweig and Hillel 2008). This may be evident in large countries, where different geographical regions and climatic zones can exhibit compensatory responses to ENSO events. Case in point: While the large 1997/98 El Niño *cost* the US economy $4 billion in losses, at least one assessment also found that it *benefited* the economy by $19 billion; yielding a $15 billion net economic gain (Changnon 1999).

The final point to be made is the difficulty of classifying super or extreme ENSO events—and any causal links to climate change—based on what has inarguably been a limited amount of records when considered the against historical and geological timescales. Moreover, historical ENSO events examined across the last millennia reveal equally extreme fluctuations can occur in the *absence* of anthropogenic forcing (Cobb et al. 2003). Indications of 'mega-El Niño', operating on 300 to 500 year cycles, further signifies that there remains more to understand (Meggers 1994). Unsurprisingly, current assessments remain somewhat divided on ENSO projections on account of climate change (Wang et al. 2017). It is in this context that the most recent IPCC report (AR 5) placed *low confidence* on exactly how ENSO will unfold under anthropogenic forcing (IPCC 2013). In summary, like some other facets of climate change, uncertainty persists in understanding exactly how it will influence ENSO across the coming century.

Despite these challenges, the 'precautionary principle' may be usefully invoked as a reason to drive greater understanding of future extreme ENSO events. There are three important aspects to this. Firstly, current research indicates that climate change is poised to increase the frequency of extreme El Niño events. Cai et al. (2014) have modeled that under worst case emissions scenarios extreme El Niño will *double* in occurrence in the twenty-first century. In this scenario, extreme El Niño will occur *once every 10 years*, but with no identified increase in intensity. Secondly, similar climate modelling for La Niña events also show a doubling in the frequency of extreme La Niña events: from one in every 23 years to *one in every 13 years* (Cai et al. 2015).

The crucial finding here, and of particular interest from a security perspective, is that approximately 75% of extreme La Niña events occur *in the year following* an extreme El Niño event. This could have major consequences for food, economic and human security, as well as the stability of nations—particularly those that are highly vulnerable and unable to respond to one, let alone consecutive, extreme climatic events. It also has significant meaning for first responders, particularly institutions such as military forces and emergency services (police/fire/ambulance) but also for aid organisations, mobilising and deploying for reconstruction and recovery. In this case, the prospect of simultaneous (or near-simultaneous) extreme ENSO events is likely to stretch the capacity of these organisations to effectively respond.

The third (and final) point to be made is that, independent of ENSO considerations, climate change is radically altering the background conditions such that even 'normal' conditions may wreak widespread destruction. To take one example, the oceans have steadily warmed such that average sea surface temperatures have increased by 0.11 °C per decade since the 1970s (IPCC 2013). The consequences of this not only causes damage in its own way (mass coral bleaching, destruction of fisheries habitats, species movement and so on) but becomes an enabler to extreme El Niño conditions. It becomes, in one sense, what climatologists might call a 'positive feedback'.

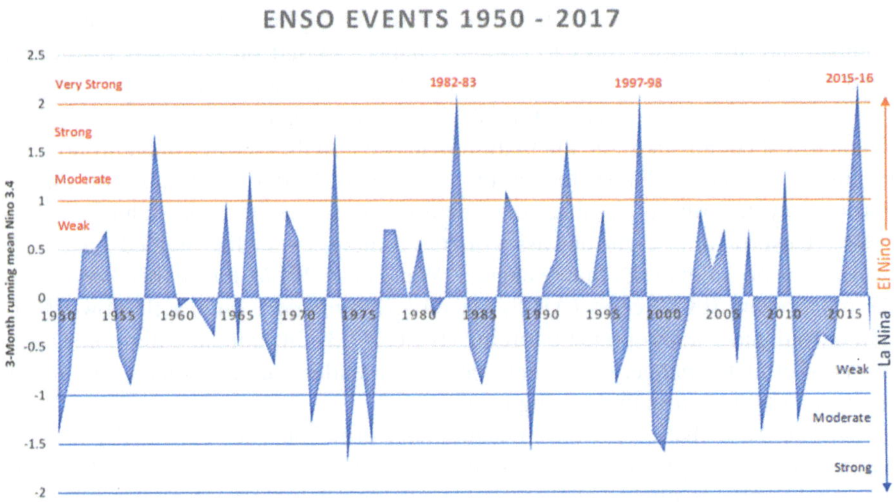

Fig. 13.3 El Niño and La Niña index, 1950–2016. *Source*: NOAA (2017a)

4 The Security Teleconnections of Extreme ENSO Events

This section examines the climate-security teleconnections of three extreme ENSO events including 1982/83 (El Niño), 1997/98 (El Niño), and 2015/16 (El Niño). As a proxy for 'climate (in)security', an examination of *deaths* caused by *drought* and *wildfire* (1950–2016) was made for eleven regions using data obtained from The International Disaster Database from the Centre for Research on the Epidemiology of Disasters (CRED). Regions included Africa, Asia, the Americas and Oceania. These regions were selected since this is where El Niño can be found to have a measurable impact on climatic conditions. Furthermore (and despite the limitations identified below), drought can be considered a reasonable proxy for climate (in) security since it is an important factor in food security (through its contribution to crop destruction), water security (via reduced rainfall), economic security (through its contribution to reducing agricultural exports and consequent loss of livelihood) and social unrest caused by increases in the cost of staple foods.

Results in Appendix 1 show that between the years 1951–2016 (a total of 66 years) some 33 years experienced El Niño conditions, that is half of all years were El Niño, as designated by the Australian Bureau of Meteorology. Six years were considered in this analysis as being 'super' or 'extreme' events; 1982–83, 1997–98 and in 2015–16 (Fig. 13.3).

Table 13.1 shows the total number of deaths for drought and wildfire by region between 1951–2016 was roughly 2.21 million people. Of this, almost 95% of deaths occurred during an El Niño year. Of the six extreme El Niño years, drought and wildfire were responsible for roughly 452,000 deaths (about 20% of all deaths). Nine of the eleven regions experienced more than 50% of deaths by drought and wildfire in El Niño years. Only people in Australia & New Zealand, and Middle

Table 13.1 Deaths by drought and wildfire in regions. *Source*: CRED (2017)

	AFRICA					ASIA - PACIFIC				AMERICAS		
	Northern Africa	Southern Africa	East Africa	West Africa	Middle Africa	South East Asia	Southern Asia	Aus/NZ	Micronesia Melanesia Polynesia	North America	South America	Total Deaths
Total deaths due to Drought & Wildfire	150,077	630	543,610	22	3070	9677	1,500,637	981	84	180	147	2,209,115
Total deaths in El Nino years	150,077	561	422,884	17	12	9342	1,500,381	278	84	101	87	2083,24
Total deaths in non El Nino years	0	66	120,726	5	3058	335	256	703	0	79	60	125,288
Total deaths in Super El Nino years	150,047	525	300,000	15	3	1222	25	81	84	30	32	452,064
As a percentage deaths in Super El Nino years	99.98%	83%	55%	68%	0%	13%	0.00%	8%	100%	17%	22%	20%
As a percentage deaths in El Nino years	100.00%	89%	78%	77%	0%	97%	99.98%	28%	100%	56%	59%	94%

Africa were less likely to die from drought or wildfire during an El Niño year. This was somewhat surprising, particularly given the large impact that El Niño has on these two regions. A quick assessment, however, shows that the situation merits further analysis. Australia, for example, was dominated by the 1967 drought that, according to the CRED database, claimed 600 lives, and the 2009 Black Saturday bushfire that claimed 190 lives. But both of these events occurred at the end of an extended dry period that included significant El Niño events (although in this analysis the 1967 deaths did not fall within an El Niño year). The CRED database indicates 600 lives were lost in the 1967 drought, but this number requires verification. The author found deaths attributable to the 1967 Tasmanian bushfires (62 lives) but was unable to find evidence of wider loss of lives due to the drought or other fires (see for example: Australian Government 2017).

Noteworthy from Table 13.1 was the high proportion of death due to drought and wildfire during El Niño years in the Pacific (100%), Northern Africa (100%), Southern Africa (89%), Southern Asia (99%), and South East Asia (97%). Death by drought and wildfire during extreme El Niño years was more likely to occur in five of the eleven regions (East Africa, Southern Africa, Northern Africa, Western Africa, and the Pacific region). All deaths by drought and wildfire in the Pacific region (defined here as Melanesia, Micronesia and Polynesia) occurred in an extreme El Niño year. (Of these, the 1997/98 extreme El Niño saw more than 200 ADF personnel deployed to assist with drought relief in PNG across September 1997 to May 1998). Other high proportions were Northern Africa (99%) and Southern Africa (83%). Conversely, death by drought and wildfire were less likely to occur in extreme El Niño years in the regions of South East Asia, South Asia, Australia & New Zealand, North America, South America and Middle Africa.

What broad conclusions can be drawn? First, if the findings from Cai et al. (2014) are to be heeded, then the burden (in terms of lives lost) of increasing frequency of extreme El Niño events will likely be experienced by some of the world's poorest regions, which are least able to deal with the consequences of climate change (i.e. African and Pacific nations). This observation reinforces other findings that climate change produces an unequal distribution of impacts (Roberts and Parks 2007; Dennig et al. 2015). The second observation is that in the study period examined, half of those years endured an El Niño event but were responsible for around 94% of deaths by drought and wildfire. More worryingly perhaps, extreme El Niño years accounted for less than one-tenth of the study period but accounted for about *one-fifth* of all deaths. Serious effort and consideration, therefore, must be made by security planners to better understand what the implications will be for a planet where El Niño becomes the dominant ENSO state and where extreme El Niño events are more common. Drought, wildfire, destroyed crops, food price hikes, loss of livelihood, mass internal displacement, and migration are all very real consequences that have significant implications for human, national and international security.

The other key finding is that *all* deaths attributable to drought and wildfire in the Pacific occurred during super El Niño years. Thus, if super El Niños are to become more frequent and more intense, then—in the absence of effective adaptation—it might be anticipated that Pacific countries will continue to disproportionately suffer the consequences of a changing climate. This is an important finding, not just for

Australian Defence officials planning future humanitarian relief missions, but also for the diplomatic efforts, its regional missions, and NGOs when developing adaptation and aid-relief strategies. Relatedly, Defence planners and aid officials must reappraise how these potential changes may impact on issues of force structure, training, and capability mix. If humanitarian and disaster relief missions are to become more frequent, for instance, then this may portend a requirement to adjust these aspects accordingly. More broadly, it may also require new regional military-to-military and military-to-civilian arrangements that go beyond current set-piece planning and to envision a much more environmentally disrupted region.

Lastly is the realisation that the ENSO phenomena—via the oscillation of energy in oceanic and atmospheric systems across the equatorial Pacific—produce global effects that significantly impact human endeavour. This chapter has shown how death by drought and wildfire are more likely to occur in years of El Niño and extreme El Niño. Drought, by account of its far-reaching and negative impact on human livelihoods, can be a harbinger of insecurity. Moreover, the concept of climate-security teleconnections—introduced in this chapter—is useful in understanding how climate change can produce security consequences at great distance.

5 Limitations and Opportunity for Further Research

Notwithstanding these findings, this chapter should not be considered definitive nor comprehensive. It contains significant limitations which merit further inquiry. First are the shortfalls of relying on a single database, in this instance the International Disaster Database from the Centre for Research on the Epidemiology of Disasters (CRED). This research would benefit from using multiple sources, including other international disaster databases (e.g. the World Food Program reports on numbers of people requiring food assistance), government statistics, journal articles, newspaper articles and so on. These additional sources could be used to verify and to triangulate a far more comprehensive and granular understanding of drought events.

Secondly, these results can be somewhat misleading insofar that drought can be a notoriously difficult disaster to define and quantify. This is mainly due to there being many contributing factors, not least of all poor governance, corruption, economic mismanagement, and other such pre-existing conditions (including armed conflict) that tend to reinforce or exacerbate the deleterious effects of El Niño. The North Korean famine of the mid 1990s that killed some half-million people is one such example where natural events (in the forms of floods and drought) were compounded by human-agency such as the end of subsidies from the former Soviet Union, the decline of incentives to produce, and the collapse of the DPRK food production system (Goodkind et al. 2011).

Thirdly, this research could be strengthened by more precisely identifying the dates for El Niño against the actual date of death for those cited as casualties of drought and wildfire. In this regard, this brief study does not account for, or closely consider, what might well be expected: that the impact of a climate anomaly can often have a delayed effect.

6 Conclusion

This chapter has examined the climate-security teleconnections of extreme ENSO events. A climate-teleconnection occurs when oceanic and atmospheric interactions produce climatological or meteorological effects across widely separated regions. From this, climate-security teleconnections may arise when an extreme climatic phenomena or climate anomaly in one geographical location can be a cause-for, or catalyst-of, the unleashing of societal disruption, instability, violence or conflict in a distant part of the planet. Drought is one example where a climate-teleconnection can produce adverse security consequences. Not least of all is the impact on human security, where this study showed that almost 95% of deaths by drought and wildfire occurred in an El Niño year, and where some 20% of all deaths happened during an extreme or 'super' El Nino year. These are important findings that should prompt Australian foreign policy and security planners to better understand the nature of climate security risks.

Appendix 1 (Table 13.2)

Table 13.2 Number of deaths by drought and wildfire, 1951–2016. Source: Data on Disasters from CRED (2017); ENSO SOI data from BoM (2017b)

Year	Jan	Feb	Mar	Apr	May	Jun	Jul	Aug	Sep	Oct	Nov	Dec	East Africa	Southern Africa	South East Asia	Southern Asia	Aus/NZ	Micronesia Melanesia Polynesia	Middle Africa	Northern Africa	Western Africa	South America	North America
1951	16.5	9.6	-1.4	-1.3	-6.6	5	-8.2	-0.5	-7	-8	-3.4	-3											
1952	-9.2	-7.9	0.2	-8.8	6	7.4	3.5	-3.7	-3.4	1.8	-0.7	-12.6											
1953	2.2	-6	-5.8	-0.5	-31.9	-2.3	-1	-17.2	-13	-0.1	-2	-4											
1954	6	-3.6	-0.9	6.9	4.4	-1.5	4.2	10.4	4.5	1.8	3.9	12.8											
1955	-5.4	15.2	2.9	-3	13.1	16.4	19.2	14.9	14.1	15.2	15.1	9.3											
1956	11.3	12.4	9.4	11.1	17.9	12.3	12.6	11	0.2	18.3	1.9	10.3											11
1957	5.6	-2.2	-0.9	1.2	-12.2	-2.3	0.9	-9.5	-10.6	-1.3	-11.9	-8.5											
1958	-16.8	-6.9	-1.4	1.2	-8.2	0.2	2.2	7.8	-3.4	-1.9	-4.7	-6.5											
1959	-8.7	-14	8.4	3.6	2.8	-6.3	-5	-5	0.2	4.2	11.1	8.2											
1960	0.3	-2.2	5.6	7.8	5.2	-2.3	4.8	6.6	6.9	-0.7	7.2	6.7											
1961	-2.5	6.3	-20.9	9.4	1.3	-3.1	2.2	0.1	0.8	-5	7.2	13.8											
1962	17	5.3	-1.4	1.2	12.3	5	-0.4	4.6	5.1	10.3	5.2	0.6											
1963	9.4	3	7.3	6.1	2.8	-9.6	-1	-2.4	-5.2	-12.9	-9.5	-11.6											
1964	-4	-0.3	8.4	13.5	2.8	7.4	6.8	14.3	14.1	12.8	2.6	-3	50										
1965	-4	1.6	2.9	-12.9	-0.3	-12.8	-22.6	-11.4	-14.2	-11.1	-17.9	1.6	2000			1500000							
1966	-12	-4.1	-13.9	-7.1	-9	1	-1	4	-2.2	-2.5	-0.1	-4			8000								
1967	14.6	12.9	7.8	-3	-3.5	6.6	1.6	5.9	5.1	-0.1	-4	-5.5					662						
1968	4.1	9.6	-3	-3	14.7	12.3	7.4	0.1	-2.8	-1.9	-3.4	2.1					12						
1969	-13.3	-6.9	1.8	-8.8	-6.6	-0.6	-6.9	-4.4	-10.6	-11.7	-0.1	3.7											
1970	-10.1	-10.7	1.8	-4.6	2.1	9.9	-5.6	4	12.9	10.3	19.7	17.4											
1971	2.7	15.7	19.2	22.6	9.2	2.6	1.6	14.9	15.9	17.7	7.2	2.1											
1972	3.7	8.2	2.4	-5.5	-16.1	-12	-18.6	-8.9	-14.6	-11.1	-3.4	-12.1											
1973	-3	-13.5	0.8	-2.1	2.8	12.3	6.1	12.3	13.5	9.7	31.6	16.9	100000										
1974	20.8	16.2	20.3	11.1	10.7	2.6	12	6.6	12.3	8.5	-1.4	-0.9	19000										
1975	-4.9	5.3	11.6	14.4	6	15.5	21.1	20.7	22.5	17.7	13.8	19.5											
1976	11.8	12.9	13.2	1.2	2.1	0.2	-12.8	-12.1	-13	3	9.8	-3											
1977	-4	7.7	-9.5	-9.6	-11.4	-17.7	-14.7	-12.1	-9.4	-12.9	-14.6	-10.6					3						
1978	-3	-24.4	-5.8	-7.9	16.3	5.8	6.1	1.4	0.8	-6.2	-2	-0.9			63						2		
1979	-4	6.7	-3	-5.5	3.6	5.8	-8.2	-5	1.4	-2.5	-4.7	-7.5				18							
1980	3.2	1.1	-8.5	-12.9	-3.5	-4.7	-1.7	1.4	-5.2	-1.9	-3.4	-0.9											
1981	2.7	-3.2	-16.6	-5.5	7.6	11.5	9.4	5.9	7.5	-5	2.6	4.7	100000		8				3000				
1982	9.4	0.6	2.4	-3.8	-8.2	-20.1	-19.3	-23.6	-21.4	-20.2	-31.1	-21.5			280								
1983	-30.6	-33.3	-28	-17	6	-3.1	-7.6	0.1	9.9	4.2	-0.7	0.1	300000	500			75			150000		20	
1984	1.3	5.8	-5.8	2	-0.3	-8.7	2.2	2.7	2	-5	3.9	-1.4			230								
1985	-3.5	6.7	-2	14.4	2.8	-9.6	-2.3	8.5	0.2	-5.6	-1.4	2.1	1017		84		11				4		
1986	8	-10.7	0.8	1.2	-6.6	10.7	2.2	-7.6	-5.2	6.1	-13.9	-13.6	200										
1987	-6.3	-12.6	-16.6	-24.4	-21.6	-20.1	-18.6	-14	-11.2	-5.6	-1.4	-4.5	237			300							
1988	-1.1	-5	2.4	-1.3	10	-3.9	11.3	14.9	20.1	14.6	21	10.8											1
1989	13.2	9.1	6.7	21	14.7	7.4	9.4	-6.3	5.7	7.3	-2	-5											
1990	-1.1	-17.3	-8.5	-0.5	13.1	1	5.5	-5	-7.6	1.8	-5.3	-2.4											
1991	5.1	0.6	-10.6	-12.9	-19.3	-5.5	-1.7	-7.6	-16.6	-12.9	-7.3	-16.7		5	57								28
1992	-25.4	-9.3	-24.2	-18.7	0.5	-12.8	-6.9	1.4	0.8	-17.2	-7.3	-5.5				56							
1993	-8.2	-7.9	-8.5	-21.1	-8.2	-16	-10.8	-14	-7.6	-13.5	0.6	1.6											3
1994	-1.6	0.6	-10.6	-22.8	-13	-10.4	-18	-17.2	-17.2	-14.1	-7.3	-11.6					4			22		24	13
1995	-4	-2.7	3.5	-16.2	-9	-1.5	4.2	0.8	3.2	-1.3	1.3	-5.5											
1996	8.4	1.1	6.2	7.8	1.3	13.9	6.8	4.6	6.9	4.2	-0.1	7.2											1
1997	4.1	13.3	-8.5	-16.2	-22.4	-24.1	-9.5	-19.8	-14.8	-17.8	-15.2	-9.1					2	60				10	
1998	-23.5	-19.2	-28.5	-24.4	0.5	9.9	14.6	9.8	11.1	10.9	12.5	13.3		25	912					47	15	1	2
1999	15.6	8.6	8.9	18.5	1.3	1	4.8	2.1	-0.4	9.1	13.1	12.8	206	4	13	181	1					15	1
2000	5.1	12.9	9.4	16.8	3.6	-5.5	-3.7	5.3	9.9	9.7	22.4	7.7	21			57						8	14
2001	8.4	11.9	6.7	0.3	-9	1.8	-3.7	-8.2	1.4	-1.9	7.2	-9.1		31					58				2
2002	2.7	7.7	-5.2	-3.8	-14.5	-6.3	-7.6	-14.6	-8.2	-7.4	-6	-10.6	588	3			2		1				
2003	-2	-7.4	-6.8	-5.5	-7.4	-12	2.9	-1.8	-2.2	-1.9	-3.4	9.3	9				4						19
2004	-11.6	9.1	0.2	-15.4	13.1	-15.2	-6.9	-7.6	-2.8	-3.7	-8.6	-8	80										
2005	1.8	-28.6	0.2	-11.2	-14.5	2.6	0.9	-6.9	3.9	10.9	-2	0.1	149				12						5
2006	12.7	0.1	13.8	14.4	-9.8	-6.3	-7.6	-15.9	-5.8	-16	-1.4	-3.5					4						5
2007	-7.8	-2.7	-1.4	-3	-2.7	5	-5	2.7	1.4	5.4	9.2	14.4		28						8		8	8
2008	14.1	21.3	12.2	4.5	-3.5	4.2	2.2	9.1	13.5	13.4	17.1	13.3	53	34								4	1
2009	9.4	14.8	0.2	8.6	-7.4	-2.3	1.6	-5	3.9	-14.7	-6	-7					180						2
2010	-10.1	-14.5	-10.6	15.2	10	1.8	20.5	18.8	24.9	18.3	16.4	27.1	20000						8			31	
2011	19.9	22.3	21.4	25.1	2.1	0.2	10.7	2.1	11.7	7.3	13.8	23										1	9
2012	9.4	2.5	2.9	-7.1	-2.7	-10.4	-1.7	-5	2.6	2.4	3.9	-6										12	3
2013	-1.1	-3.6	10.5	0.3	8.4	13.9	8.1	-0.5	3.9	-1.9	9.2	0.6					3				1		22
2014	12.2	-1.3	-13.3	8.6	4.4	-1.5	-3	-11.4	-7.6	-8	-10	-5.5					2					12	2
2015	-7.8	0.6	-11.2	-3.8	-13.7	-12	-14.7	-19.8	-17.8	-20.2	-5.3	-9.1			30		2	24	3			1	12
2016	-19.7	-19.7	-4.7	-22	2.8	5.8	4.2	5.3	13.5	-4.3	-0.7	2.6				25	2						16
2017	1.3	-2.2	5.1	-6.3	0.5	-10.4	8.1	3.3															
												Grand Total	543,610	630	9,677	1,500,687	981	84	3,070	150,077	22	147	180
													East Africa	Southern Africa	South East Asia	Southern Asia	Aus/NZ	Micronesia Melanesia Polynesia	Middle Africa	Northern Africa	Western Africa	South America	North America

El Nino Year
Super El Nino

References

Ashok, K., Zhaoyong, G., Saji, N. H., & Yamagata, T. (2004). Individual and Combined Influences of ENSO and the Indian Ocean Dipole on the Indian Summer Monsoon. *Journal of Climate, 17*(16), 3141–3155. https://doi.org/10.1175/1520-0442(2004)017<3141:iacioe>2.0.co;2.

Australian Government. (2017). Natural Disasters in Australia. Commonwealth of Australia, Accessed 10 Sept. http://www.australia.gov.au/about-australia/australian-story/natural-disasters.

BoM. (2017a). Southern Oscillation Index. Commonwealth of Australia, Accessed 31 May 2017. http://www.bom.gov.au/climate/glossary/soi.shtml.

BoM. (2017b). Southern Oscillation Index (SOI) since 1876. Bureau of Meteorology, Accessed 10 Sept. http://www.bom.gov.au/climate/current/soihtm1.shtml.

Brunner, A. D. (2002). El Niño and world primary commodity prices: Warm water or Hot air? *The Review of Economics and Statistics, 84*(1), 176–183. https://doi.org/10.1162/003465302317332008.

Cai, W., Borlace, S., Lengaigne, M., Van Rensch, P., Collins, M., Vecchi, G., Timmermann, A., Santoso, A., McPhaden, M. J., & Wu, L. (2014). Increasing frequency of extreme El Nino events due to greenhouse warming. *Nature Climate Change, 4*(2), 111.

Cai, W., Wang, G., Santoso, A., McPhaden, M. J., Wu, L., Jin, F.-F., Timmermann, A., Collins, M., Vecchi, G., Lengaigne, M., England, M. H., Dommenget, D., Takahashi, K., & Guilyardi, E. (2015). Increased frequency of extreme La Niña events under greenhouse warming. *Nature Climate Change, 5*, 132–137. https://doi.org/10.1038/nclimate2492.

Changnon, S. A. (1999). Impacts of 1997—98 El Niño generated weather in the United States. *Bulletin of the American Meteorological Society, 80*(9), 1819–1827. https://doi.org/10.1175/1520-0477(1999)080<1819:ioenog>2.0.co;2.

Cobb, K. M., Charles, C. D., Cheng, H., & Lawrence Edwards, R. (2003). El Nino/Southern oscillation and tropical pacific climate during the last millennium. *Nature, 424*(6946), 271–276. http://www.nature.com/nature/journal/v424/n6946/suppinfo/nature01779_S1.html.

CRED. (2017). Centre for Research on the Epidemiology of Disasters – CRED. Edited by School of Public Health Université catholique de Louvain. Belgium.

Dennig, F., Budolfson, M. B., Fleurbaey, M., Siebert, A., & Socolow, R. H. (2015). Inequality, climate impacts on the future poor, and carbon prices. *Proceedings of the National Academy of Sciences, 112*(52), 15827–15832. https://doi.org/10.1073/pnas.1513967112.

Dijk, v., Albert, I. J. M., Beck, H. E., Crosbie, R. S., de Jeu, R. A. M., Yi, Y. L., Podger, G. M., Timbal, B., & Viney, N. R. (2013). The millennium drought in southeast Australia (2001–2009): Natural and human causes and implications for water resources, ecosystems, economy, and society. *Water Resources Research, 49*(2), 1040–1057. https://doi.org/10.1002/wrcr.20123.

Goodkind, D., Loraine W., & Peter J. (2011). *A reassessment of mortality in North Korea, 1993–2008*. Population Association of America Washington, DC, 28 March 2011.

Grant, P. R., Rosemary Grant, B., Keller, L. F., & Petren, K. (2000). Effects of El Nino Events on Darwin's Finch Productivity. *Ecology, 81*(9), 2442–2457. https://doi.org/10.1890/0012-9658(2000)081[2442:EOENOE]2.0.CO;2.

Holbrook, N. J., Davidson, J. L., Feng, M., Hobday, A. J., Lough, J. M., McGregor, S., & Risbey, J. S. (2009). El niño-southern oscillation. In E. S. Poloczanska, A. J. Hobday, & A. J. Richardson (Eds.), *A Marine climate change impacts and adaptation report card for Australia 2009*. NCCARF.

Hong, Li-Ciao. 2016. *Super El Niño*, Springer Theses. Singapore: Springer.

IPCC. (2013). Summary for Policymakers. In T. F. Stocker, D. Qin, G.-K. Plattner, M. Tignor, S. K. Allen, J. Boschung, A. Nauels, Y. Xia, V. Bex, & P. M. Midgley (Eds.), *Climate change 2013: The physical science basis. Contribution of working group I to the fifth assessment report of the intergovernmental panel on climate change*. Cambridge, UK and New York, NY: Cambridge University Press.

Laosuthi, T., & Selover, D. D. (2007). Does El Nino affect business cycles? *Eastern Economic Journal, 33*(1), 21–42.

Lee, T., & McPhaden, M. J. (2010). Increasing intensity of El Niño in the central-equatorial Pacific. *Geophysical Research Letters, 37*(14), n/a–n/a. https://doi.org/10.1029/2010GL044007.

Lim, Y.-K., & Schubert, S. D. (2011). The impact of ENSO and the Arctic Oscillation on winter temperature extremes in the southeast United States. *Geophysical Research Letters, 38*(15), n/a–n/a. https://doi.org/10.1029/2011GL048283.

Lindsey, R. (2016). Global impacts of El Niño and La Niña. NOAA, Last Modified 9 February 2016, Accessed 4 June. https://www.climate.gov/news-features/featured-images/global-impacts-el-ni%C3%B1o-and-la-ni%C3%B1a.

Liu, J., Yuan, X., Rind, D., & Martinson, D. G. (2002). Mechanism study of the ENSO and southern high latitude climate teleconnections. *Geophysical Research Letters, 29*(14), 24–1-24-4. https://doi.org/10.1029/2002GL015143.

Meggers, B. J. (1994). Archeological evidence for the impact of mega-Niño events on Amazonia during the past two millennia. *Climate Change, 28*(4), 321–338. https://doi.org/10.1007/BF01104077.

NOAA. (2017a). ENSO 3-Month running mean. NOAA National Weather Service Climate Prediction Center, Accessed 3 Sept http://www.cpc.ncep.noaa.gov/products/analysis_monitoring/ensostuff/ensoyears.shtml.

NOAA. (2017b). Equatorial Pacific sea surface temperatures. NOAA, Accessed 31 May. https://www.ncdc.noaa.gov/teleconnections/enso/indicators/sst.php.

NOAA. (2017c). Niño Regions. National oceanic and atmospheric administration, Accessed 3 June https://www.ncdc.noaa.gov/teleconnections/enso/indicators/sst.php.

Peterson, R. G., & White, W. B. (1998). Slow oceanic teleconnections linking the Antarctic Circumpolar Wave with the tropical El Niño-Southern oscillation. *Journal of Geophysical Research: Oceans, 103*(C11), 24573–24583. https://doi.org/10.1029/98JC01947.

Plisnier, P. D., Serneels, S., & Lambin, E. F. (2000). Impact of ENSO on East African ecosystems: a multivariate analysis based on climate and remote sensing data. *Global Ecology and Biogeography, 9*(6), 481–497. https://doi.org/10.1046/j.1365-2699.2000.00208.x.

Roberts, J. T., & Parks, B. C. (2007). *A climate of injustice: global inequality and climate change–vulnerability; responsibility and action.* Cambridge: MIT Press.

Rosenzweig, C., & Hillel, D. (2008). *Climate variability and the global harvest impacts of El Niño and other oscillations on Agro-ecosystems.* Oxford: Oxford University Press.

Stott, L., Poulsen, C., Lund, S., & Thunell, R. (2002). Super ENSO and Global climate oscillations at millennial time scales. *Science, 297*(5579), 222–226. https://doi.org/10.1126/science.1071627.

Taylor, A. H., Jordan, M. B., & Stephens, J. A. (1998). Gulf Stream shifts following ENSO events. *Nature, 393*(6686), 638–638.

Timmermann, A., Oberhuber, J., Bacher, A., Esch, M., Latif, M., & Roeckner, E. (1999). Increased El Nino frequency in a climate model forced by future greenhouse warming. *Nature, 398*(6729), 694–697.

Trenberth, K. E., & Hoar, T. J. (1997). El Niño and climate change. *Geophysical Research Letters, 24*(23), 3057–3060. https://doi.org/10.1029/97GL03092.

Trenberth, Kevin. 2016. "What North America can expect from El Niño." *The Conversation*, accessed 6 June. http://theconversation.com/what-north-america-can-expect-from-el-nino-51959.

Ubilava, D. (2012). El Niño, La Niña, and world coffee price dynamics. *Agricultural Economics, 43*(1), 17–26. https://doi.org/10.1111/j.1574-0862.2011.00562.x.

Wang, C., Deser, C., Yu, J.-Y., DiNezio, P., & Clement, A. (2017). El Niño and Southern Oscillation (ENSO): A Review. In P. W. Glynn, D. P. Manzello, & I. C. Enochs (Eds.), *Coral reefs of the eastern tropical Pacific: Persistence and loss in a dynamic environment* (pp. 85–106). Dordrecht: Springer.

WHO. (2015). El Niño Southern Oscillation (ENSO) and health. United Nations World Health Organisation, Accessed 12 June. http://www.who.int/globalchange/publications/factsheets/el-nino-and-health/en/.

Yuan, Y., & Yang, S. (2012). Impacts of different types of El Niño on the east asian climate: Focus on ENSO cycles. *Journal of Climate, 25*(21), 7702–7722. https://doi.org/10.1175/jcli-d-11-00576.1.

Chapter 14
From Skin to Strategy: Repositioning the Terrain and Temporality of Heat Stress through Social Practices

Elspeth Oppermann and Matt Brearley

1 Introduction

Heat stress, the net heat load to which an individual is exposed, has been a topic of military research for well over half a century (Yaglou and Minard 1957). Defence forces have been at the forefront of understanding the heat stress risks posed by hot and humid conditions. To date, heat stress has been approached largely as a technical and medical problem, usually as an effect of exposure to, and exertion in, hot environments. Typically, it is managed through behavioural protocols that govern rates of exertion, physiological acclimatization programs, technological innovations to cool bodies and hydration/nutrition interventions. The most intensive research has been focused on combat personnel where material and technological solutions have been more thoroughly explored (O'Hara et al. 2008). There are also general provisions in place for work health and safety in broader operational and support roles (see for example DoD 2014).

In these respects, typical defence force strategies for handling heat stress are two-fold. The first strategy is focused on short-term exercises or operations for personnel on active duty. Although modified over time, the same basic formulation has been in place since the 1950s (d'Ambrosio Alfano et al. 2014), where Wet Bulb Globe Temperatures, factoring in temperature, humidity and solar radiation, have been used to determine levels of exertion and the ratio of work to rest. The second strategy promotes acclimatisation to new environments through programs that gradually increase exposure to, and exertion in, the new conditions.

This chapter draws on emergent, cross-disciplinary research that combines thermal physiology with Social Practice Theory to posit an alternative understanding of

E. Oppermann (✉) · M. Brearley
Heat Stress Research Partnership, Northern Institute, Charles Darwin University, Darwin, NT, Australia
e-mail: Elspeth.oppermann@cdu.edu.au

S. Pearson et al. (eds.), *Australian Contributions to Strategic and Military Geography*, Advances in Military Geosciences,
https://doi.org/10.1007/978-3-319-73408-8_14

heat stress as a hazard and its management. By doing so, we also contribute to the growing literature on embodied and sensory geographies (see for example Pink 2015; Paterson 2009; Hitchings 2007), which include the analysis of military activity (Gregory 2015; Gregory 2016). In relation to contemporary military practice, our approach allows us to make two related conceptual critiques about how heat stress is construed spatially and temporally. The first is that, in terms of heat stress, the environment or terrain of operation is co-constituted by the operation itself, requiring greater attention to how the materiality and temporality of heat stress is understood. We then go on to examine the implications of this critique at tactical, operational and strategic levels.

2 Landscape or Skin-Scape: Reimagining Co-Produced Thermal Terrains

The military is one of the most advanced sectors in the management of heat stress, with a long history of identifying it as a problem and producing technologies to manage it. However, by identifying environmental conditions as a limit on the amount of exertion personnel may undertake, heat is seen as an external problem that impinges on operational capacity. This can result in heat management being ignored in favour of achieving operational objectives. While there is limited public evidence of this (see, for an indirect example, Miller et al. 2011), internationally, heat illness is known to be heavily under-reported (Stacey et al. 2016) and there is an observable gap between 'best-practice' heat management and actual practices reported during military heat fatality investigations (Rav-Acha et al. 2004; Casa et al. 2012). One response to such outcomes is the argument that physiological indicators need to be improved in order to inform how force protection is balanced with operational objectives (Hunt et al. 2016). However, the deep tensions that arise when heat management is characterised as an obstacle to operational objectives may also play a role. This oppositional representation obscures the intimate way in which heat stress emerges through operational activity, and its significance in enabling or constraining the achievement of set objectives.

We propose an alternative framing of the problem based on thermal physiology and informed by Social Practice Theory (SPT). SPT is one stream of the emerging scholarship on embodied ways of knowing and learning, and there are multiple theories of social practice. We draw on Schatzki (2010) who accounts for social practices as 'what make sense' for people to do as they engage in everyday activities. This account of social practice is given further analytical traction through the claim that practices are co-constituted by meanings, materials and competences (Shove 2003; Shove et al. 2012). *Meaning* comprises the knowledges and values that are called upon to justify or explain practices; *materials* are the physical tools, fabrics, technologies or capabilities that are required for—and deployed in—conducting a practice; and *competences* are the embodied skills and abilities required to conduct the practice. Social Practices refer to both the particular performance of a practice and to the generalised performances of that practice. To make this distinction, we refer to Practice-as-Entity, an example of which would be 'running' as an abstract noun. Particular performances of this practice are known as Practice-as-Performance, such as 'a run' or 'running' as a verb.

In military terms, performing a practice, such as conducting a particular activity, might be made up of: certain meanings, such as specific operational objectives, broader doctrine, rules and protocols; materials, such as environmental conditions, physical fitness, equipment, technical capabilities and the affordances of particular materials; and embodied competences, such as situational awareness and physical skills such as loading a weapon. Particular meanings, materials and competences are drawn together to constitute and perform each practice, or enact the activity the practice enables.

Thermal physiology considers heat stress as emerging from exposure to environmental conditions as well as exertion. However, exertion and exposure are in turn an outcome of particular performances of a Practice (Oppermann et al. 2017b; Oppermann and Walker in press). That is, while heat stress may be a physiological 'problem', it becomes one as a result of how a human body is involved in activities that cause it to be exposed, to exert itself, to hydrate, to rest, or to actively cool itself - all of which are socially mediated.

Social Practice readings of heat include the argument of Shove and colleagues that heat is largely an incidental effect of other practices such as exercise (Shove and Walker 2014; Shove et al. 2014; Walker et al. 2014). While this is true, heat management is also an explicit practice. In the military it often is, as attested by the range of heat management strategies indicated in the introduction. The conceptual tension here gives us some insight into why heat management can be so problematic, as both positions are accurate; heat is incidental to many practices (such as operational activities) and it can also be deliberately shaped through explicit heat stress management practices. In situations where heat stress is an issue, particularly where its management appears to disrupt desirable practices (such as work, a sports match or a military activity), the failure to consider both the incidental and deliberate production and management of heat together may result in ineffective heat management at best and dangerous responses at worst.

How we think of heat emerging 'together' appears more clearly if we think of heat stress as emerging at the interface of activity and environment (articulated through practice). Such an epistemic position is informed also by socio-ecological approaches at the routinized, entity 'end' of the practice heuristic. Significantly for understanding Practice-as-Performance, there has been rapid development over the last decade of methodologies designed to interrogate the embodied relationship between activity and environment, including but not limited to: embodied geography and sensory ethnography, drawing on a consideration of the haptic, somatic and affective co-production of practice, particularly in relation to the experiences of bodies in place (Stanes and Gibson 2017; Waitt and Stanes 2015; Jerstad 2016). Heat stress is an excellent topic for thinking about how materiality is not only something that the human is in-relation with, but as something humans are intimately co-constituted by (Oppermann and Walker in press).

On this basis, if heat stress occurs as a result of exertion and exposure in an environment, then the terrain producing heat and heat stress as a hazard is not just the physical, external environment, but is rather co-produced by how the body interacts with that environment. We are not talking about simply autonomic responses. This is also distinct from merely identifying risk factors or vulnerabilities that predispose personnel to heat stress. Rather, taking on Social Practice Theory, it identifies the contingency of why practices are constituted in the way that they are, and therefore how they contribute to heat stress and how this might be avoided. The shift is slight, but useful; it allows for the consideration of preventative measures as well as responsive ones, and also of wider sources of the hazard over which personnel have more control (than the weather).

This means that the 'terrain' of a military activity or operation – physiologically speaking at least - is co-constituted by the performance of military practices themselves. This can be thought of as a 'thermal geography' that takes place not only in landscapes and climatic zones, but across the more minute, micro-temporal, micro-spatial (including micro-material) body-environment nexus: materialities such as ambient temperature, water vapour pressure, and wind, but also the more intimate stocks and flows of thermal energy via contact with materials such as skin and clothing (Oppermann and Walker in press). Thermodynamics and thermal physiology meet in a complex system of stocks and flows of energy and matter between bodies and their environment, down to the 'skin-scape' of the number of sweat glands exposed to the air, to wind, or where skin comes into contact with different kinds of substances (such as metal, plastic, clothing, rock) with different conductive or convective affordances. Bodies are responsive to these micro-environments but this same responsiveness means the interface between body and environment is an actively co-produced terrain. As much as mountains or communications networks, skin-scapes and broader thermal geographies shape military capability and capacity, for example by enabling or limiting physical and cognitive performance.

Such an account of terrain has resonances with the work of a number of critical geographers (Dalby 2013b; Steinberg and Peters 2015; Clark 2015) who have explored terrain as not being flat or two-dimensional but rather 'volumetric' (Elden 2013) - incorporating air and space, and constituted by tangible materialities. Just how tangible these materialities are is well illustrated in Santanu Das' (2005) work on the 'slimescapes' of the First World War. Here, the terrain is far from inert; the front-line landscape of mud and slime actively intervened in the conflict, not only impeding soldiers' movement, but affecting their perceptions and emotions. The profound somatic relationship between environment and bodies could also be read as one of the co-constituted 'natures of war' (Gregory 2016). For Derek Gregory (2015), the embodiment of war in soldiers' bodies, as much as in the landscape, was related to the emergence of strong ways of knowing the environment though non-visual senses of touch, smell and sound, which he called a 'corpography'. The sensorial intensity and sense of physical contingency present in Das and Gregory's accounts has some similarities to Australian analyses of the experiences of persistent environmental heat and chronic heat stress which co-constitute geographies of activity while also being rendered 'visible' through a wider use of physical senses (de Vet 2017; Oppermann et al. 2015).

This broader conceptualisation of terrain, as both active and sense-able, is also in keeping with shifts in military theory and its critiques (Dillon 2007; Dillon and Reid 2009). The shift in question is a move to a dynamic account of the 'terrain of operation' as the always-emergent convergence of physical environment, information flows and technical capabilities. This is exemplified in Network Centric Warfare, initially conceptualised by US Admiral Arthur Cebrowski (Cebrowski and Garstka 1998) as understandings of informatics contingency, and was subsequently applied to knowledges, cultural development and emerging social terrains (González 2017). This shift in how war was conceptualised and practiced was taken up more broadly, including by the Australian Defence Force (O'Malley 2010). Adopting a Foucauldian analytical lexicon, network centric war is deeply connected to advanced liberal modes of governance – including through governing the contingency through which human life (understood biologically, and for our purposes, physiologically) emerges, and through governance of the self, as the internalisation of the norms, or values and mechanisms through which life (also understood socially) emerges (O'Malley 2004).

In the same way that these accounts demonstrate military capability and capacity is contingent on an array of socio-political as well as technological components, and how these are arranged and related informatically, heat stress can similarly be understood as something that contingently emerges through the temporal, spatial and material imbrications of components and how this co-production is managed. This alternative positioning means that, rather than seeing heat management as disruptive of military activities, activities all inherently produce and affect heat stress. While this may seem to increase the problem that heat stress poses, it in fact opens up the times, spaces and ways in which heat and heat stress can be managed. Rather than seeing heat management as an external inhibitor, heat production/loss considerations can be integrated into how activities are performed, in terms of choice of tactics, operational design and strategic decisions, so as to enhance their effectiveness.

Utility of course does not entail ethics. Foucault's famous inversion of Clausewitz's dictum, from 'war is politics by other means' to 'politics is war by other means', has a deep relevance here. From assigning to individuals the responsibility for managing their own body heat and behaviours, to the wider strategic politics of heat engendered by climate change policy, heat and its management are inherently political. Insofar as for Foucault war and politics are modes through which freedom was subdued, heat management is also an ethical problem. Any intervention into how heat is managed also attempts to shape practices, and therefore seeks to create particular subjectivities and ways of life. The precise operation of heat management at the bodily level, including through bodily discipline and surveillance (Pink and Fors 2017) and enrolment in technologies that 'responsibilise' individuals for its successful management, begs the question of what politics it enacts directly. Ensuring the resilience of military personnel in hostile conditions is one such politics (O'Malley 2010), laudable for operational purposes but an additional weight and potential danger for the individual. Indirectly, the production of these resilient subjects enables other forms of power such as military capacity, and other politics such as liberal capitalism. How adept subjects are at managing heat also enables society more broadly to cope with increasingly hot conditions, as well as the extent to which certain local or global temperature thresholds are seen as critical, thereby driving (or not) a politics of greenhouse gas mitigation (Dalby 2013). For these reasons, rendering heat stress and heat management visible through attention to the practices that produce heat and heat stress (or do not), the practices that enact its management (or not), and identifying the characteristics and objectives of these practices, also provides a more graduated platform for the consideration of politics and ethics, regardless of the type of military activity in which they occur.

3 From Terrain to Temporality

Within the military, heat has typically been positioned as an 'acute' or exceptional problem, occurring in unfamiliar conditions, on particularly hot days or during particularly laborious tasks (Oppermann et al. 2017a). If we were to apply the classical 'levels of war' here, heat, heat stress and its management would be confined to the tactical level, as not only a small scale but also a short term phenomenon. This imaginary is challenged in persistently hot regions where heat stress becomes a chronic, rather than acute, problem. The global tropics, including Australia's monsoon tropics, experiences extended periods of high temperatures and very high humidity, creating a long-term environmental basis for heat stress (Oppermann et al. 2017b). However, as a result of climate change, the increase in the frequency, severity and duration of heat waves in sub-tropical, temperate and arid regions of Australia means that these regions may also experience a shift in the temporality (the occurrence, frequency and duration) of these hazards. Rather than being 'exceptional' events, heat waves are becoming part of the character of the local climate, resulting in broad and lasting effects across environmental, social, political and economic systems (Bolitho and Miller 2017; Hughes et al. 2016).

Extreme heat has been a highly significant natural hazard in Australia, killing more people than all other natural hazards combined since colonization (Coates et al. 2014). While extreme heat generally does not require a military response, it does shape the conditions in which military operations take place. From a Social Practices perspective, environmental conditions are implicit in all activities as allowing or disallowing, enabling or constraining the effective and safe operation of human systems, infrastructure and materials (Oppermann and Walker in press). As such, while they have micro-spatial and micro-temporal significance, they also play out more broadly at the operational and strategic level. To demonstrate this, the next section explores how heat and heat stress can be produced and managed at multiple scales. We have used the tactical, operational and strategic categorisations of 'levels of war' as a heuristic to structure the discussion, but we recognise that not all military activities are directly war-related and that the boundaries between the tactical, operational and strategic are not always clear-cut as military activities can have effects at more than one 'level'. To briefly exemplify our argument below, we indicate that heat stress too plays out through inherently contingent temporalities and spatialities across the emergent terrain of military activity and capability (MacGregor 1992; Dunn 1996), far beyond its presumed 'acute' and tactical location.

4 Tactics

One current definition of the tactical level of warfare is: "battles and engagements [that] are planned and executed to achieve military objectives assigned to tactical units or task forces" (DoD 2017:232). This immediate, localised scale was historically the domain of the soldier or unit, where body heat was seen as the problem to be managed and where failing to do so successfully could affect the success of the activity. Military history is punctuated by examples of the ineffective management of body heat. An early example was provided in the year 400 Before the Common Era (BCE) approximately, by Athenian and Spartan troops who were worn out by 'thirst and sun' (Goldman 2002). Later in 24 BCE, the Roman military expedition into Arabia was decimated by heat stroke (Jarcho 1967). Approximately 19 centuries later, similar issues were reported for the British military deployed to India with troops also affected by heat stroke (Hunter 1887). More recently (1980–2002), the US Army had in excess of 5000 hospitalisations and 37 deaths due to heat illness over a 22-year period (Carter III et al. 2005).

In the Australian Defence Force, heat is seen as a problem at the tactical level in two related ways: First as an environmental problem, through exposure to the local climate, weather conditions, or the immediate environment (such as tarmac, equipment, machinery). Second, as a problem arising from exertion as the internal production of heat, particularly when the external environment is unforgiving. Understood as short-term, localized problems of bodies or their immediate activities and environments, heat is also managed in a similar fashion - through short term, temporary solutions (Oppermann et al. 2017a). These include rescheduling of activities to avoid the harshest conditions, application of engineering controls such as shielding from radiant heat, prescribed work to rest ratios based on environmental conditions, cooling methods to accelerate dissipation of body heat, and medical plans that include transporting heat affected soldiers out of hot environs (Department of Defence 2014; Department of Defence 2016a).

The management of heat in tactical activities can be understood through Social Practice Theory. If we recall that social practices are constituted of materials, meaning and competence, this helps make explicit the elements through which heat is co-produced. Many military protocols implicitly recognise these elements. For example, work/rest cycles acknowledge that military activities shape personnel's exposure to hot environments, and the level of exposure entailed by these activities is inherently related to the kinds of tactic being used. Furthermore, it is implicitly acknowledged that the nature of the activity (its intensity and duration) will entail particular work rates, which are more difficult to sustain in higher temperatures. As such, tactic choice co-produces how hot bodies become in conjunction with the environment. How and when tactical activities are conducted (for example, in the middle of the day or at night, with or without machinery or equipment to reduce exertion) actively shapes bodily heat production, and consequently heat stress. This relocates heat from being an external hazard managed by internal mitigation measures, to being an effect of activity in an environment. While the shift is subtle, the location of heat at the interface of activity and environment allows the nature of those activities to come into question, allowing attention in turn to how these are constituted. This allows heat to be considered as a systemic property of military activity rather than as an externality that is disruptive of planned activity.

Social Practice theory is well placed to understand how activities are performed in relation to environments. In each performance of a practice military personnel try to carry out its activity in terms of 'what makes sense' (Schatzki 2010) in the situation, given the meanings, materials and competences that are available (or not). This means that tactical activities are often implicitly—but can be explicitly—shaped in relation to: the knowledge, behaviours and values of personnel and their units; the materiality of the environment, individual bodies (such as levels of hydration) and their equipment, including clothing, cooling and associated technologies; and the competences of personnel, such as individuals' embodied awareness of their thermal tolerance, ability to continue exerting themselves or acknowledge heat stress symptoms.

The analysis of practice elements is of course not confined to considering heat alone. In fact, heat may be incidental to the practice. For example, the knowledge that conducting an operation at night provides a tactical advantage, having the materials or technology to enable it (such as night-vision goggles) and the training to conduct such activities appropriately, incidentally confers significant reduction of heat exposure in hot environments by operating at cooler times. Building heat considerations into tactical decision-making would need to be supported through considerations of the requisite knowledges (and objectives), materials (and technologies) and (embodied) competences. This conceptual orientation may already be implicitly practiced by professionals encountering these environments on an everyday basis, but has yet to be explicitly theorised in military doctrine and military geography, let alone mainstreamed into official protocol.

5 Operations

Operations are considered to be the "level of warfare at which campaigns and major operations are planned, conducted and sustained to achieve strategic objectives within theatres or other operational areas" (DoD 2017:178). It is well established that there is a scalar, systemic relationship between tactics, operations and strategy. As the authors have examined elsewhere, there is also a relationship between the micro-scale and macro-scale management and effects of heat (Oppermann et al. 2017a). Chronic 'everyday' exposure, while affecting the efficacy of tactics, can also play out at an operational level. An example is the reduced effectiveness resulting from multiple and subsequent engagements over an extended period of time, wherein consistent heat stress related sleep deprivation might affect vigilance and decision making at tactical to strategic levels (Miller et al. 2011; Lieberman et al. 2005; Johnson and Kobrick 2001). On the other hand, the increased thermal tolerance of chronically acclimatized personnel can also result in improved physical capacity (Brearley et al. 2016). In theory, at least, this can contribute to better tactical outcomes which, particularly if sustained, would support better operational outcomes.

Hot environmental conditions and the ability to manage heat may also impact military operations indirectly, through affecting civilian systems. Examples include the production of food and other goods, and the provision of services and infrastructures (Oppermann et al. 2017a). The likelihood of heat stress being significant in these systems was previously discounted on the basis that societies were assumed to be more-or-less adapted to their thermal environments. However, climate change impacts, conflict and natural disasters all affect the ability of extant socio-technical systems to handle environmental extremes. A pertinent example of this is the conclusion of the Australian Department of Defence that the projected increases in the frequency and magnitude of severe weather events will extend the operational season and level of demand on the Australian Defence Force (Department of Defence 2016b; Press et al. 2013).

In relation to heat specifically, the increased frequency and intensity of heat waves and rising global temperatures means that the effects of heat stress at the tactical level will similarly increase. However, increases in frequency, duration and intensity could have cumulative, knock-on-effects, and even direct effects at an operational scale. Such issues are also faced when, more simply, conducting an operation in a hotter climatic zone than usual, raising questions such as: what might consistent exposure mean for where an operation's personnel are sourced from, in terms of ensuring they are better physically acclimatized (materials), have the necessary embodied capabilities (competences) or climatically appropriate knowledge (meaning) to successfully incorporate heat management into tactical decisions and operational design? When would it be best to schedule operations and how might changing seasons or climate change affect the design of operations? How might Rest and Recreation scheduling change in terms of location, duration, and frequency? What kind of technology or infrastructure investment might reduce the impost of sustained operations in regularly or consistently hot environments? In Australia's monsoon tropics, heat (including the effects of humidity) is raising such questions (Oppermann et al. 2017a) and appears to have been considered in the Australian Defence Force's decisions to schedule large-scale training exercises during the cooler months.

6 Strategy

Strategy can be defined as the "level of warfare at which a nation, often as a member of a group of nations, determines national or multinational (alliance or coalition) strategic security objectives and guidance, then develops and uses national resources to achieve those objectives" (DoD 2017:226). The discussion so far has demonstrated that the impost of heat on individual bodies can cascade into tactical effects that can, in turn, accumulate into reduced effectiveness at an operational level. Systemic considerations of operational viability in strategy are important too. There are also basic thermal considerations than can directly impact the formulation of strategic objectives. One example is reconsidering the strategic value of certain locations on account of their changing environmental conditions. What makes a location geostrategic is most often related to trade routes and transport systems, the protection of national territories or facilities, and the capabilities and capacities of defence platforms and networks. However, the functioning of these systems and the utility of these locations is in part shaped by heat (with melting arctic sea ice and changing shipping lanes a key example), and also, we would argue, by heat stress. Under future conditions, it is projected that human settlements in certain areas will become near to impossible to sustain due to heat stress (Kjellstrom et al. 2016). Even in less affected areas, heat stress will combine with other climate change impacts, including the associated toll of extreme heat on crops, livestock and food and water distribution systems (World Bank 2012) to affect the value and viability of regions. Depending on the objectives and values that shape the temporal horizon of military strategy and how explicitly this is connected to wider systemic vulnerabilities, heat and heat stress may or may not be incorporated into strategic decisions, but it will affect strategic outcomes, regardless.

One of the uses of climate change projections in military strategy has been to assess the changing risks to major facilities, (Brzoska 2015; Busby 2008), but increased exposure of their personnel to extreme heat may also affect their strategic value in the short, medium or long-term. US naval facilities in Florida, as well as military facilities in Guam and Diego Garcia, are at risk from intensifying hurricanes and rising seas (Busby 2008). While the impacts of climate change-related sea level rise have been broadly considered, the exacerbating effect of climate change on the duration and magnitude of extreme heat, while considered in relation to productivity and public health (see for example UNDP 2016) has not received the same attention in military strategic-level responses to the concomitant changes in strategic military geography. Similarly, while there has been a raging debate about the links between climate change, drought and conflict, the role of heat and heat stress in the liveability of a region, its economic viability and indeed levels of violence (Mueller et al. 2014; Zander et al. 2016; Simister and Cooper 2005; McDermott et al. 2017) has received less attention as a strategic issue. As we have argued elsewhere in relation to Australia's monsoon tropics (Oppermann et al. 2017a), heat impacts on the viability of military establishments, although less direct or visible than sea-level rise, is likely to be significant. Therefore, the duration and level of projected hot conditions ought to be a factor in determining investment in new Defence training facilities where there are likely to be numerous personnel and/or equipment with operating thresholds likely to be exceeded in hotter (including more humid) conditions.

In Australia, the idea that future climate could shape the prospect of maintaining a population in its tropical monsoon region seems uncannily reminiscent of the early twentieth century environmental determinism of Griff Taylor (O'Brien 2015). However, if social practices (including military practices and procedures) shape the production and management of heat stress, then how these are constituted—in relation to technological, cultural, economic and political innovations and considerations—will enable or disable particular strategic objectives or modes of securing them. For example, the cost of operating in some parts of Australia as a result of heat stress on humans and materials may shift the way in which strategic objectives are pursued. This could include utilising more automation to reduce physical exertion in hot regions, or moving away from heavy personnel concentrations altogether to unmanned or remote surveillance and strike capability (Oppermann et al. 2017a).

Projected temperature rises and the associated impacts of heat stress also shapes short term, but no less strategic, objectives. The United Nations Framework Convention on Climate Change (UNFCCC) process, including more recently the Paris Agreement, are examples. These activities demonstrate yet another profound way in which Social Practices actively constitute terrains of military activity – both through shaping the magnitude of emissions and therefore anthropogenic warming of the planet, and the constitution of the regulatory and political frameworks that shape military action.

Preparations for, and responses to, these impacts play out at all scales: at national and international scales through political and economic agreements; at regional scales where changing and expanding thermal geographies of risk are observable; and, at the scale of human bodies and lived experience, impacts on health, wellbeing and productivity (Oppermann et al. 2017b; Turton 2017). In turn, these impacts scale-up into significant regional and national problems, including national productivity (Zander et al. 2015) as well as recruitment and retention in particular regions (Zander et al. 2016; Zoellner et al. 2017). As a result of these scalar effects, heat stress may also be a significant source of economic and political risk at the strategic level. For example, heat stress may play a role in global migration, but it has also been argued that the single biggest economic impact of climate change will be reduced productivity, particularly in the global topics, which includes Australia's hot and humid monsoon tropics and the Southeast Asian region (Kjellstrom et al. 2009). Extreme heat and heat stress, directly as well as by reducing productivity, are likely to undermine economic resilience and contribute to increased migration and general instability across the global tropics.

At a strategic level, an account of heat stress informed by Social Practice Theory produces a broader understanding of where heat stress emerges and where it can be mitigated and managed. Practice approaches can also broaden modes of intervention and provide greater flexibility and responsiveness (Oppermann et al. 2017b; Strengers and Maller 2012; Strengers and Maller 2015). However, limits to human capacity to modulate practices (in terms of their materials, meaning and competences) may mean that heat stress, particularly in the context of extreme climate change, may be most effectively and cost-efficiently managed if engaged with as a matter of strategy, particularly where the strategic extends to the 'political level' at which non-military mechanisms, such as international agreements and regulations, may serve broader military interests by reducing climate change per se, and therefore reducing the degree to which climatic change drives increases in heat stress.

7 Discussion and Conclusion

This chapter has responded to two conflicting provocations, implicit in much military and strategic thinking about heat stress and climate change. The first is that heat stress is typically characterised as an acute or exceptional event, and therefore primarily as a tactical problem. The second is that, climate change is viewed as long-term and large scale, and therefore frequently assumed to be a strategic problem (typically for long-lived assets). However, as the issue of heat stress demonstrates—particularly when viewed in terms of social practices—climate change, and hot conditions more generally, have impacts that play out at the tactical, operational and strategic scale. The extent of heat stress, its causes and opportunities for management, is more visible if heat stress is considered as an effect of how military activities are conducted in environments. Social Practice Theory provides a useful approach to analysing this co-production, through its attention to the interrelation of materials, meanings and competence. Bringing these aspects to light as contingent elements of practices helps to unpick the productive role that military activities (at all levels) play in shaping not only when and how heat stress occurs, but how it can be more effectively managed, across multiple temporal and spatial scales.

In reconceptualising heat and heat stress and their relation to military theory, we have taken a journey from skin to strategy; traversing the micro-temporal and micro-spatial aspects of the bodily regulation of heat to the macro spaces and times of regional shifts in extreme heat and associated changes in practices and facilities, to global climatic change and international agreements. In doing so we have indicated the scalar effects heat stress can have at different levels of military activity, each with their own implicit spatial and temporal characteristics, from the tactical to the strategic.

This chapter's contribution has been two-fold. First, in relation to the practical problem of how to prevent and manage heat stress, we have conceptually extended the spaces, times and ways in which heat can be managed beyond the current focus on the acute and tactical. We hope this serves as a provocation to academics and military personnel alike to consider heat stress and its management in a much broader way at all levels of military activity. Second, by analysing heat stress through a Social Practice lens we contribute a richer understanding of the 'terrain of operation' of military practices to include body-environment relations per se. In particular, we are able to identify the ways in which meanings, materials and competences come together to co-produce the emergence (or not) and management (or not) of heat and heat stress.

Insofar as Social Practice theory is an approach located in Critical and Human Geography, we also hope to have demonstrated the applicability of these less familiar sub-disciplines to Military Geography and its contributions to strategic thought. Beyond military contexts, the growing body of research in Australia that is informed by Social Practice Theory, Human and Critical Geography, as well as Embodied Geography and Sensory Ethnography, are reflective of modes of inquiry that can yield academically rich and practically useful ways of conceptualising and grappling with critical geographic problems, including in relation to heat and heat stress. Finally, by providing a platform for considering heat stress and its management in social terms, we also open up the possibility of paying more attention to the politics and ethics of the ways that military activities intervene in, enable, constrain and shape the management of heat, and, in the process, shape identity and ways of life. While a Social Practices approach might be critiqued for positioning the meanings and bodies of personnel and populations as the target of more extensive scrutiny and possible intervention, the governance of military and health practices has been doing this implicitly for a long time. Examining the political and ethical implications of technological, thermal physiology and organisational approaches currently in place will be better enabled if their effects on the registers of meaning, materials and competence can be made explicit, and therefore open to consideration and critique.

References

d'Ambrosio Alfano, F. R., Malchaire, J., Palella, B. I., & Riccio, G. (2014). WBGT index revisited after 60 years of use. *Annals of Occupational Hygiene, 58*(8), 955–970.

Bolitho, A., & Miller, F. (2017). Heat as emergency, heat as chronic stress: Policy and institutional responses to vulnerability to extreme heat. *Local Environment, 22*, 682–698.

Brearley, M., Norton, I., Rush, D., et al. (2016). Influence of chronic heat acclimatisation on occupational thermal strain in tropical field conditions. *Journal of Occupational & Environmental Medicine, 58*, 1250–1256.

Brzoska, M. (2015). Climate change and military planning. *International Journal of Climate Change Strategies and Management, 7*, 172–190.

Busby, J. W. (2008). Who cares about the weather?: Climate change and US national security. *Security Studies, 17*, 468–504.

Carter, I. I. I. R., Cheuvront, S. N., Williams, J. O., et al. (2005). Epidemiology of hospitalizations and deaths from heat illness in soldiers. *Medicine and Science in Sports and Exercise, 37*, 1338–1344.

Casa, D. J., Armstrong, L. E., Kenny, G. P., et al. (2012). Exertional heat stroke: New concepts regarding cause and care. *Current Sports Medicine Reports, 11*, 115–123.

Cebrowski, A. K., & Garstka, J. J. (1998). Network-centric warfare: Its origin and future. *US Naval Institute Proceedings*. pp 28–35.

Clark, N. (2015). Fiery arts: Pyrotechnology and the political aesthetics of the Anthropocene. *GeoHumanities, 1*(2), 266–284.

Coates, L., Haynes, K., O'Brien, J., McAneney, J., & de Oliveira, F. D. (2014). Exploring 167 years of vulnerability: An examination of extreme heat events in Australia 1844–2010. *Environmental Science & Policy, 42*(0), 33–44. https://doi.org/10.1016/j.envsci.2014.05.003.
Dalby, S. (2013). Biopolitics and climate security in the Anthropocene. *Geoforum, 49*, 184–192.
Das, S. (2005). *Touch and intimacy in first world war literature*. Cambridge: Cambridge University Press.
Dillon, M. (2007). Governing through contingency: The security of biopolitical governance. *Political Geography, 26*, 41–47.
Dillon, M., & Reid, J. (2009). *The liberal way of war: Killing to make life live*. London: Routledge.
DoD. (2014). *Defence work health and safety manual*. Canberra: Department of Defence.
DoD. (2016a). *Defence health manual (DHM)*. Canberra: Department of Defence.
DoD. (2016b). *Defence white paper*. Canberra: Department of Defence.
DoD. (2017). Department of Defense Dictionary of Military and associated terms. June 2017 ed. Online.
Dunn, M. (1996). Levels of war: Just a set of labels? *Research and Analysis: Newsletter of the Directorate of Army Research and Analysis* 10:4–10.
Elden, S. (2013). Secure the volume: Vertical geopolitics and the depth of power. *Political Geography, 34*, 35–51.
Goldman, R. (2002). Introduction to heat related problems in military operations. In K. Pandolf & R. Burr (Eds.), *Medical aspects of harsh environments* (pp. 3–42). Washington DC: Department of the Army.
González, R. J. (2017). *Ethnographic intelligence: The human terrain system and its enduring legacy* (pp. 51–73). Springer: Reconfiguring Intervention.
Gregory, D. (2015). Gabriel's map: Cartography and Corpography in modern war. In P. Meusburger, D. Gregory, & L. Suarsana (Eds.), *Geographies of knowledge and power*. Dordrecht: Springer.
Gregory, D. (2016). The natures of war. *Antipode, 48*, 3–56.
Hitchings, R. (2007). Geographies of embodied outdoor experience and the arrival of the patio heater. *Area, 39*, 340–348.
Hughes, L., Hanna, E., & Fenwick, J. (2016). *The silent killer: Climate change and the health impacts of extreme heat*. Canberra: Climate Council of Australia.
Hunt, A. P., Billing, D. C., Patterson, M. J., et al. (2016). Heat strain during military training activities: The dilemma of balancing force protection and operational capability. *Temperature, 3*, 307–317.
Hunter, G. (1887). Notes on heat and "heat-stroke" at Assouan in the summer of 1886. *British Medical Journal, 2*.
Jarcho, S. (1967). A Roman experience with heat stroke in 24BC. *Bulletin of New York Academy of Medicine, 43*, 767–768.
Jerstad, H. (2016). Mundane energies: The working body as a heat source in the Indian Himalayas. *Anthropology Today, 32*, 7–10.
Johnson, R. F., & Kobrick, J. L. (2001). *Psychological aspects of military performance in hot environments* (p. 26). Natick: Army Research Institute of Environmental Medicine.
Kjellstrom, T., Kovats, R. S., Lloyd, S. J., et al. (2009). The direct impact of climate change on regional labor productivity. *Archives of Environmental & Occupational Health, 64*, 217–227.
Kjellstrom, T., Briggs, D., Freyberg, C., et al. (2016). Heat, human performance, and occupational health: A key issue for the assessment of global climate change impacts. *Annual Review of Public Health, 37*, 97–112.
Lieberman, H. R., Bathalon, G. P., Falco, C. M., et al. (2005). Severe decrements in cognition function and mood induced by sleep loss, heat, dehydration, and undernutrition during simulated combat. *Biological Psychiatry, 57*, 422–429.
MacGregor, D. A. (1992). Future battle: The merging levels of war. *Parameters, 22*, 33.
McDermott, K. M., Brearley, M. B., Hudson, S. M., et al. (2017). Characteristics of trauma mortality in the Northern Territory, Australia. *Injury Epidemiology, 4*, 15.

Miller, N. L., Shattuck, L. G., & Matsangas, P. (2011). Sleep and fatigue issues in continuous operations: A survey of U.S. Army officers. *Behavioral Sleep Medicine, 9*, 53–65.
Mueller, V., Gray, C., & Kosec, K. (2014). Heat stress increases long-term human migration in rural Pakistan. *Nature Climate Change, 4*, 182–185.
O'Brien, C. (2015). Griffith Taylor's Climographs and contemporary blind spots. *Learning Communities: Objects of Governance, 15*, 26–31.
O'Hara, R., Eveland, E., Fortuna, S., et al. (2008). Current and future cooling technologies used in preventing heat illness and improving work capacity for battlefield soldiers: Review of the literature. *Military Medicine, 173*, 653–657.
O'Malley, P. (2004). *Risk, uncertainty and government*. London: Routledge Cavendish.
O'Malley, P. (2010). Resilient subjects: Uncertainty, warfare and liberalism. *Economy and Society, 39*, 488–509.
Oppermann, E., & Walker, G. (in press). Immersed in thermal flows: Heat as productive of and produced by social practices. In Y. Strengers & C. Maller (Eds.), *Social practices and more-than-humans: Nature, materials and technologies*.
Oppermann, E., Spencer, M., & Brearley, M. (2015). Emotional athletes, brainy workers and other hot new developments. *Learning Communities: International Journal of Learning in Social Contexts, 15*, 32–39.
Oppermann, E., Brearley, M., Carter, S., et al. (2017a). *Scoping study: Heat stress in Australia's monsoon tropics. Final report for the Department of Defence. Heat stress research partnership: Northern institute* (pp. 1–69). Darwin: Charles Darwin University.
Oppermann, E., Brearley, M., & Law, L. et al (2017b). Heat, health, and humidity in Australia's monsoon tropics: A critical review of the problematization of 'heat' in a changing climate. *Wiley Interdisciplinary Reviews: Climate change*: e468–n/a.
Paterson, M. (2009). Haptic geographies: Ethnography, haptic knowledges and sensuous dispositions. *Progress in Human Geography, 33*.
Pink, S. (2015). *Doing sensory ethnography*. London: Sage.
Pink, S., & Fors, V. (2017). Being in a mediated world: Self-tracking and the mind–body–environment. *Cultural Geographies, 0*, 1474474016684127.
Press, A., Bergin, A., & Garnsey, E. (2013). Heavy weather: Climate and the Australian defence force. Special Report, Australian Strategic Policy Institute, Canberra, Australia.
Rav-Acha, M., Hadad, E., Heled, Y., et al. (2004). Fatal exertional heat stroke: A case series. *The American Journal of the Medical Sciences, 328*, 84–87.
Schatzki, T. R. (2010). *Site of the social: A philosophical account of the constitution of social life and change*. Pennsylvania: Penn State University Press.
Shove, E. (2003). *Comfort, cleanliness and convenience: The social Organization of Normality (New technologies/New cultures)*. Oxford: Berg Publishers.
Shove, E., & Walker, G. (2014). What is energy for? Social practice and energy demand. *Theory, Culture & Society, 31*, 41–58.
Shove, E., Pantzar, M., & Watson, M. (2012). *The dynamics of social practice: Everyday life and how it changes*. London: Sage.
Shove, E., Walker, G., & Brown, S. (2014). Material culture, room temperature and the social organisation of thermal energy. *Journal of Material Culture, 19*, 113–124.
Simister, J., & Cooper, C. (2005). Thermal stress in the USA: Effects on violence and on employee behaviour. *Stress and Health, 21*, 3–15.
Stacey, M. J., Brett, S., Woods, D., et al. (2016). Case ascertainment of heat illness in the British Army: Evidence of under-reporting from analysis of medical and command notifications, 2009–2013. *Journal of the Royal Army Medical Corps, 162*, 428–433.
Stanes, E., & Gibson, C. (2017). Materials that linger: An embodied geography of polyester clothes. *Geoforum, 85*, 27–36.
Steinberg, P., & Peters, K. (2015). Wet ontologies, fluid spaces: Giving depth to volume through oceanic thinking. *Environment and Planning D: Society and Space, 33*(2), 247–264. https://doi.org/10.1068/d14148p.

Strengers, Y., & Maller, C. (2012). Materialising energy and water resources in everyday practices: Insights for securing supply systems. *Global Environmental Change, 22*, 754–763.
Strengers, Y., & Maller, C. (2015). *Social practices, intervention and sustainability: Beyond behaviour change*. London and New York: Routledge.
Turton, S. M. (2017). Expansion of the tropics: Revisiting frontiers of geographical knowledge. *Geographical Research, 55*, 3–12.
UNDP. (2016). *Climate change and labour: Impacts of heat in the workplace*. New York: United Nations Development Programme.
de Vet, E. (2017). Experiencing and responding to everyday weather in Darwin, Australia: The important role of tolerance. *Weather, Climate, and Society, 9*, 141–154.
Waitt, G., & Stanes, E. (2015). Sweating bodies: Men, masculinities, affect, emotion. *Geoforum, 59*, 30–38.
Walker, G., Shove, E., & Brown, S. (2014). How does air conditioning become 'needed'? A case study of routes, rationales and dynamics. *Energy Research & Social Science, 4*, 1–9.
World Bank. (2012). *Turn down the heat: Why a 4 degrees centigrade warmer world must be avoided*. Washington DC, USA: The World Bank.
Yaglou C.P. and Minard D. (1957).Control of Heat Casualties at Military Training Centers. *American Medical Association Archives of Industrial Health. 16*(4): 302–316.
Zander, K., Oppermann, E., & Garnett, S. (2015). Extreme heat poses a billion-dollar threat to Australia's economy. *The Conversation*. Online at: https://theconversation.com/extreme-heat-poses-a-billion-dollar-threat-to-australias-economy-41153 [Accessed 10 Oct 2017].
Zander, K. K., Surjan, A., & Garnett, S. T. (2016). Exploring the effect of heat on stated intentions to move. *Climatic Change, 138*, 297–308.
Zoellner, D., Brearley, M., & Oppermann, E. (2017). Regional disparities in apprentice attrition rates: Heat and quarter four's significance in northern Australia. *International Journal of Training Research, 15*(2), 179–195.

Chapter 15
Sustainably Managing the Defence Estate: Selected Case Studies

Richard Thackway and Stuart Pearson

1 Introduction

Military geography provides opportunities for systematically describing and understanding the complex of biophysical and human systems that create contested landscapes of value for military and other purposes (Woodward 2004). Defence training areas are valued for specific military purposes and yet there are other consequential values related to prior use and subsequent uses that military geography can address. The Australian Defence Organisation (Defence) states that its vision for the Defence environmental policy has four pillars: compliance, efficiency, trust and accountability (DoD 2016). The transparent way it manages the sites and environmental factors to sustain the Defence estate, manage impacts, minimise pollution and recognise heritage is held as an exemplar. The results, tools and systems continue to improve (albeit from a low base in the 1960s), driven by community expectations and the requirements of the Environment Protection and Biodiversity Conservation Act 1999 (EPBC Act), the Defence Estate Quality Management System (DEQMS) and Defence's Environmental Strategy and plans (DoD 2016). These were recently refreshed in developing environmental factor management within a specific Garrison and Estate Management System (GEMS) (DoD 2017).

Standing on those pillars and systems are the informed decision makers who purposefully manage geographically specific sites for various purposes. Australia's Defence estate—of about 3.5 million ha (0.5% of the Australian land area)—is an asset managed for Defence purposes as a part of a mosaic of publicly and privately managed landscapes. The Defence estate is widely spread across nine out the ten of Australia's agro-climatic regions (Thackway and Ford 2017). Most of the estate

R. Thackway (✉) · S. Pearson
School of Physical and Environmental Science, University of New South Wales, Australian Defence Force Academy, Canberra, ACT, Australia

S. Pearson et al. (eds.), *Australian Contributions to Strategic and Military Geography*, Advances in Military Geosciences,
https://doi.org/10.1007/978-3-319-73408-8_15

retains its native vegetation extent, which provides various benefits including cover for Defence training activities, the protection of ecosystem services (Yapp et al. 2010), buffers to adjoining uses and additional uses such as the maintenance of biodiversity. Only a very small portion of the estate is intensively managed, that is where the vegetation is either removed and replaced with managed cover (e.g. base training areas and airfield aprons) or removed and replaced with non-vegetated ground cover (e.g. fixed target or impact areas, infrastructure and facilities).

Almost all of Australia prior to the 1850's, including Defence training areas, was managed for other resource management purposes by Aborigines (Gammage 2011) and, since then, by European colonial settlers who transformed the native ecosystems for forestry, farming and grazing. Each site has an environmental history, prior to acquisition and management for Defence purposes, that had modified the function, structure and composition of the plant communities. Over time the Defence land management regimes have also affected the condition of these estates by influencing complex ecological processes and their effects on landscape degradation, restoration and regeneration.

Depending on the specific Defence facilities and uses of an area, spatial zone plans are sometimes used to guide how different areas are managed, using these prescribed management regimes, with the aim to achieve sustainable use and management over time. However, where plans exist, the gap between plan and operation is an important one. A key issue for Defence is to maintain the capability of an asset, i.e. its extent and condition, for Defence purposes. If that asset fails to be accessible and useable for Defence purposes, then it is important that this change of state should become immediately apparent to the decision maker. That capability requires a management regime, usually operating over longer periods, that ensures sustainable land for Defence and other purposes. Meanwhile, the rest of the landscapes are transformed by management regimes on non-Defence lands and the tools for tracking changes and trends vary widely depending on the sectoral interests but are now well-developed and widely applied (Thackway and Specht 2015; Thackway and Freudenberger 2016). There is consensus that the resilience of native plant communities subject to all uses can be described using the ecosystem components of function, structure and composition (Noss 1990). This particular research on the Defence estate sought to test if the Defence areas that are transformed, relative to a reference and/or baseline, can be measured and assessed using the open source and leading national assessment framework, Vegetation Assets States and Transitions (VAST-2) (Lesslie et al. 2010).

Decision makers seek to know the condition and the trends in asset condition at site and landscape scales. This applies equally to the Defence estate. For the Department of Defence an understanding of how the environment responds to intentional and inadvertent actions is critical for on-ground managers, policy makers and planners at local, regional and national scales. The response of the native vegetation to changes in land management regimes, which reflect social, economic and environmental values, can be seen in evidence of environmental history, geography and landscape ecology where changes in ecosystem structure, composition and function (regenerative capacity) can be assessed using standardised reproducible and robust methods.

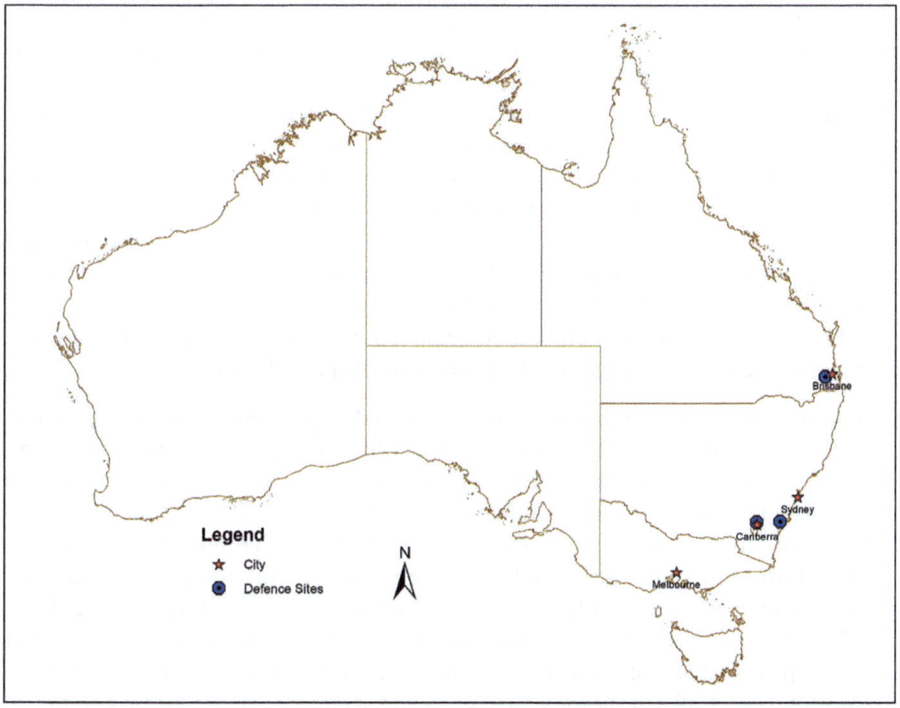

Fig. 15.1 Three case study sites selected to represent sustainable management of the Defence estate

This chapter presents the results of applying VAST-2 to transformations of plant communities using criteria and indicators of function, structure and composition at three selected sites (Fig. 15.1). Two are former Defence estates: Tianjara Defence Training Area, New South Wales (NSW), and the Belconnen Naval Transmitter Station in the Australian Capital Territory (ACT); third is the current——and rapidly developing—Royal Australian Air Force Base at Amberley in Queensland (Qld).

2 Assessment Methodology

The VAST-2 assessment framework was developed, tested and applied across Australia's 10 agro-climatic regions (Thackway and Freudenberger 2016). It was applied in various land use contexts including grazing, plantation forestry, water supply reserves, and publicly and privately protected areas (Thackway and Specht 2015). The framework has been applied in a diversity of vegetation formations, including tropical rainforests, savanna woodlands, desert shrublands, and alpine treeless grassland and herbfields (Lesslie et al. 2010; Thackway and Specht 2015;

Thackway and Freudenberger 2016) and for various land management regimes including saw log production in native eucalypt forests, production agriculture and protected area management. Publications and a web site (http://vasttransformations.com/) provide details of the method and its applications, along with case study examples from national to small sites scales (Thackway 2016).

The assessment framework guides the evaluation of what land management changes have occurred at sites over time and how these have affected 22 response indicators grouped into 10 criteria representing ecological function, structure and composition (Table 15.1). It captures the key stages of the degradation and recovery of ecosystem processes modified by human activity.

A brief overview of the methodology, applied consistently to the three Defence estate case studies to demonstrate the framework, is provided here:

- Site and areas were determined by a soil-landscape association and georeferenced as a centroid, so it is a consistent location in time, now and into the future.
- A systematic tabular chronology of the land management history and the observed responses of the 22 indicators and 10 criteria representing ecological function, structure and composition (Table 15.1) was compiled for each site since European settlement. Each site history and literature review were built using published and unpublished accounts, scientific surveys, long-term ecological monitoring sites, land manager interviews, maps and remote sensing and public-private data archives. To provide an accurate comprehensive chronology for each site, draft chronologies for each site were peer reviewed by ecologists and land managers.
- A focused effort was needed to identify and describe the unmodified or reference state plant community type for each site using the same 22 indicators and 10 criteria to provide the relative transformations of each site over time.
- These integrated site-based environmental histories show the effects that land management practices have on vegetation condition and resilience over time. An aggregate index for each year in the chronology of a site was calculated across four levels in a hierarchy (Table 15.1). The response of the plant community at each site was described in terms of land management regimes and practices (Table 15.2). This usefully shows how regimes individually and collectively transform vegetation resulting in its maintenance, enhancement, restoration, degradation and/or removal and replacement.
- It is critical to make a distinction between the reference state and a contemporary baseline. VAST-2 is better because it uses a fully-natural reference state described using criteria and indicators listed in (Table 15.1) rather than the current state or no baseline (Thackway and Freudenberger 2016). Reference states, used to assess the transformation of case study sites, were obtained from published sources or were elicited from skilled local ecologists and botanists.

It should be noted that in this assessment the methods used to populate each of the indicators were previously published for recording site-based effects of land use and land management practices on the condition of native plant communities (Thackway 2014). Fundamental to the successful application of this assessment framework is access to credible, published sources of information that describe

Table 15.1 List of Vegetation Assets States and Transition (VAST) indicators that meet criteria and reflect overall vegetation condition (further explanation Thackway 2014)

Condition Components[1]	Key Functional, Structural and Composition Criteria	Indicators
Level 3	Level 2	Level 1
Functional (55%)	Soil hydrology	Rainfall infiltration and soil water holding capacity
		Surface and subsurface flows
	Soil physical status	Effective rooting depth of the soil profile
		Bulk density of the soil through changes to soil structure or soil removal
	Soil nutrient status	Nutrient stress: rundown (deficiency) relative to reference soil fertility
		Nutrient stress: excess (toxicity) relative to reference soil fertility
	Soil biological status	Organisms maintaining soil porosity and nutrient recycling
		Surface organic matter, soil crusts
	Natural disturbance regime	Area/size of disturbance events
		Interval between disturbance events
	Reproductive potential	Reproductive potential of overstorey structuring species
		Reproductive potential of understorey structuring species
Structural (27%)	Overstorey structure	Overstorey top height (mean)
		Overstorey foliage projective cover (mean)
		Overstorey structural diversity (i.e., a diversity of age classes)
	Understorey structure	Understorey top height (mean) of the plant community
		Understorey ground cover (mean) of the plant community
		Understorey structural diversity (i.e., a diversity of age classes)
Compositional (18%)	Overstorey composition	Densities of overstorey species functional groups
		Richness: the number of indigenous overstorey species relative to the number of exotic species
	Understorey composition	Densities of understorey species functional groups
		Richness: the number of indigenous understorey species relative to the number of exotic species
Total Vegetation Status (100%) Level 4		

[1]Modified from the functional, structural and compositional levels of organization observed in biological diversity (Noss 1990)

Table 15.2 Five land management regimes outcomes used to describe effect on native plant community, relative to the reference state (Thackway and Lesslie 2006)

	Management Regimes
1	Practices that do not affect indicators of vegetation function, structure and composition
2	Practices that harvest/reduce vegetation products (biomass, fibre, flowers, fruit and nuts), which affect indicators of vegetation function, structure and composition
3	Practices that enhance or improve indicators of vegetation function, structure and composition
4	Practices that extirpate or remove indicators of the function, structure and composition
5	Practices that reconstruct or reinstate indicators of the function, structure and composition

Modified from Thackway and Freudenberger (2016)

where, when, and what changes in land use and land management practices have occurred (Table 15.2). The sources of this information are multidisciplinary and include land use histories, remote sensing, and several branches of ecological science landscape, vegetation and restoration. Much of the historic information is qualitative, in contrast to contemporary information, which is mostly quantitative. A key challenge for users of that handbook is to compile information that is generally piecemeal, not easily discoverable, poorly organised and, to a great extent, non-digital (Thackway 2014).

A simple graphical report card was prepared to show the drivers of change and trend relative to the reference state, to clearly communicate the transformation. The graph shows a trajectory for the site where the condition score is composed of a potential 100% (i.e. an unmodified reference state). The total score is comprised of three consistently weighted components: function or regenerative capacity (55% weighting); vegetation structure (27% weighting); and species composition (18% weighting). The total vegetation status score was calibrated to the six VAST classes described in Thackway and Lesslie (2006, 2008) enabling consistent description and mapping of the state changes over time across widely different vegetation types and agro-climate regions. The degree of divergence between the reference state (100%) and the vegetation scores for each case study represents the degree of modification over time. Scores were grouped according to the following intervals to provide a meaningful basis for describing and summarising change and shown, using the colours below, on the subsequent figures:

80.1%–100% of the reference state corresponds to a residual/unmodified state (Class I);

60.1%–80% corresponds to a modified state (Class II);

40.1%–60% corresponds to a transformed state (Class III);

20.1%–40% corresponds to VAST Class IV: replaced and adventive; as well as

1%–20% corresponds to VAST Class V: replaced and managed; and VAST Class VI: removed and replaced (<1%)

For example, Class IV might represent a grassland, where the biomass or ground cover is equally dominated by native and exotic herbs and grasses. Class II might represent a native woodland that was modified by a land management regime and is recovering toward to the reference state, Class I.

3 Three Case Studies

Three Defence properties were selected to represent the three Australian military services: the Royal Australian Army (Tianjara Defence Training Area, NSW), the Royal Australian Navy (Belconnen Naval Transmitter Station, ACT) and the Royal Australian Air Force (RAAF) Base Amberley, Qld (Fig. 15.1). The degree to which the three sites were transformed by military management activities, from a reference state vegetation description of the plant communities, was assessed using VAST-2.

3.1 Tianjara Defence Training Area

The former Tianjara Defence Training Area is located west of Nowra and is about 18,500 ha (Thackway 1985's, Figure 2). The main artillery impact area was focused on Mount Tianjara (4800 ha) which prior to 1968 was remote and only accessible via horseback, on-foot and off-road vehicles.

The focus for this assessment, builds on Thackway's assessment of the transformation of the dry mallee low heaths (Thackway 1985's, Figure 11) since first settlement of the region. Dry mallee low heaths are one of the most widespread plant communities on the Tianjara Plateau that repeatedly occur on uniform sandy shallow soils, gentle slopes with aspects south to west. The soils are poorly drained but dry out during the hot summer months assisted by desiccating westerly winds. The broader landscape is comprised of a complex of mallee and dry low heaths on shallow soils overlying rock platforms to tall open forests on deep humic soils and cool temperate rainforest in sheltered gullies and valleys (Gunn 1983).

The Lands of Tianjara (Gunn 1983) described the structure (regenerating from a fire 2 years prior), composition and function of vegetation in the context of land units and mapped land systems. The broad regional and state-wide context for the Tianjara's *Eucalyptus stricta* mallee-heath plant communities was reliably compared and honed to plant community descriptions in the 'Native vegetation of Illawarra Escarpment and Coastal Plain' publication (NPWS 2002).

From discussions with local graziers it is known that the plant communities of the Tianjara plateau were regularly burnt by the land managers to prepare the pasture for cattle agistment, particularly during times of drought, prior to the arrival of the Army (Sneddon 2016). In the absence of regular wildfire, mallee eucalyptus form the dominant overstorey in the dry low heaths (NPWS 2002). Three severe wildfires were recorded in the district in 1939, 1957 and 1980 (Hilder 1982; Sneddon 2016; Gunn 1983).

An ecological reconnaissance study was undertaken by CSIRO of the Tianjara Defence Training Area in the early 1980s. At that time, anecdotal evidence recorded from discussions with Defence personnel indicated that dry mallee low heaths were used by the Army for field training, artillery practice range and field manoeuvres

Fig. 15.2 Impact crater 1980 in dry mallee low heath. Photograph by Richard Thackway

because of their trafficability. Army use commenced in 1942 at Tianjara Field Firing Range and continued as an artillery practice range until the early 1980s (Sneddon 2016).

During this period, Tianjara Defence Training Area was managed for Defence purposes, including using controlled burns to assist with access and trafficability. CSIRO surveys show that by 1982, the impact of the Army's management was only slight; impact craters were very few and small (Fig. 15.2) and where present were usually filling-in by slumping earth and sheet erosion and many supported regenerating indigenous low vegetation (Gunn 1983). A detailed ecological and vegetation study was completed in the main impact area, adjacent to the southern arm of Mount Tianjara. That field study assessed the extent and condition of the plant communities and the associated bird communities and reported on the effects of military training activities (Gunn 1983; Thackway 1985's, Figure 11). In the mid-1970s (Snedden pers. comm) and early 1980s (Thackway 1985) the dry mallee heaths were typically low in height, low in cover and had minimal structural diversity because of fire management practices and a recent severe wildfire.

Public-private efforts in the late 1960s through to the late 1970s were made to transfer the tenure of Tianjara Defence Training Area from a military training area to a public recreation and conservation area. Due to public pressure to enable the wider upper Shoalhaven Catchment to be recognized and used for public recreation values, the Provisional Occupancy Lease granted to Defence was not extended beyond 1980 (Sneddon 2016).

Defence ceased managing the Tianjara Defence Training Area in the early 1980s and the area was transferred to the NSW National Parks and Wildlife Service

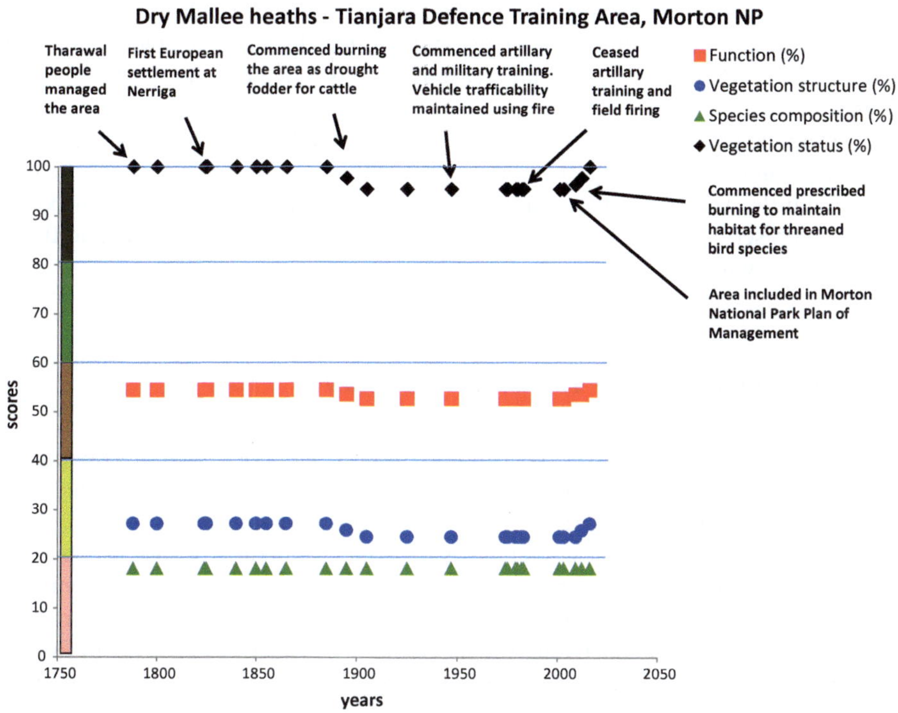

Fig. 15.3 Tianjara Defence Training Area NSW dry mallee heath site. Showing land management practices effects on the structure, composition and function/regenerative capacity of the site

(NPWS). Commencing in the early 2000s, the area was managed to prevent frequent intense wildfire burning with the intent of enhancing or improving indicators of vegetation function, structure and composition. The positive effects on vegetation condition, of controlling wildfire, commencing in 2009, are shown in Fig. 15.3.

The intent of NPWS was to enhance and maintain habitat for the Ground Parrot, Striated Fieldwren and Eastern Bristle Bird and maintain the long-term diversity and viability of the plant communities (NPWS 2002). There are ongoing concerns for the safety of walkers due to unexploded ordinance in this former Defence estate (Sneddon 2016).

Figure 15.3 records the stability of dry mallee low heaths within the impact area of the former Tianjara Defence Training Area. They were not obviously modified by Europeans due to the rugged terrain and very shallow soils with impeded drainage. Despite the area being used as an access route for overland droving of small herds of cattle to the coast and as crown leases granted for cattle grazing between 1895 to 1941, the landscape as a whole has remained unfragmented and largely unmodified (i.e. VAST Class I).

It should be noted that this area is relatively remote from the surrounding more intensively managed grazing country and was only used for drought-based adjustment of domestic stock. During drought cattle were walked to the area with horses and drovers, rather than transporting stock to the area with stock trucks. Military vehicles and training within the area also did not introduce appreciable numbers of exotic species (Gunn 1983). For this reason, there is a higher score for the species composition.

During the presence of the Army (1942–1983), and the area being used for artillery range and other training, the extent and condition of the dry mallee low heaths changed little. Transfer of the area into the Morton National Park and the introduction of a prescribed system of fire management has restored the structure and function of the dry mallee low heaths.

3.2 Belconnen Naval Transmitter Station

The former Belconnen Naval Transmitter Station was once Natural Temperate Grassland of the South Eastern Highlands of Australia. This plant community was previously widespread in the region, but is now highly fragmented. It exists regionally in several condition states: retained as remnant (Unmodified); variously altered (Modified or Transformed); converted to a mixed native/exotic pasture (Adventive); converted and managed as exotic pasture (Replaced and managed); or converted to urban areas or infrastructure (Removed and replaced) (Thackway and Lesslie 2008). These grasslands were once regionally widespread across the lowlands of the ACT, estimated to cover 20,000 ha prior to European settlement, but are now fragmented and very highly modified with approximately 5% intact. This plant community is listed in Critically Endangered Threatened Ecological Communities under the EPBC Act and temperate grasslands are described as one of the most threatened ecosystems in Australia (Hodgkinson 2014).

The former Belconnen Naval Transmitter Station is located within the suburb of Lawson, ACT and occupied approximately 143 ha (Hodgkinson 2014). The station was established in the 1940 on grazing lands. A small area of 115 ha was fenced in 1970 and that action enclosed a small population of kangaroos (Thackway 2012).

This case study shows the transformation of the temperate grassland since first settlement of the region and included the effects of grazing and fencing on the ecosystem values. The regional situation and the Territory and Commonwealth legislative protections and conservation measures for this endangered ecological community are excellently described in the report by Cooper (2009). In the grassland sites that are considered in good condition, native herbs (forbs) may comprise up to 70% of species present and the dominant stratum is between 0.25 and 1 metre tall (Hodgkinson 2014). These grasslands occur where tree growth is limited by cold air drainage, generally below 625 metres altitude. In this ecological setting, it is generally regarded that the community is naturally treeless or has less than 10% projective foliage cover of trees, shrubs and sedges in its tallest stratum. These grasslands are dominated by native species of perennial tussock grasses including *Themeda triandra*, *Austrodanthonia* species, *Austrostipa* species, *Bothriochloa macra* and *Poa* species with a relatively high diversity of native herbs (Hodgkinson 2014).

Information about how the site was managed, prior to and after the site was a Defence facility was compiled from a wide range of published and unpublished sources including government and non-government, research reports and interviews with scientists who had surveyed the site. The site was commissioned as a Defence facility in 1940 and decommissioned in 2005. The history of this site (Thackway 2012) is representative of remnant temperate grasslands in the ACT; including impacts caused by total grazing pressure, including from kangaroos, rabbits and sheep, by frequent and untimely mowing regimes and by the failure to apply an autumn fire regime, which is essential for the survival of most of the grassland plants (Hodgkinson 2014). Day-to-day management of the site is now the responsibility of the ACT Government.

In 1940, the Transmitter Station was established within a rural setting where sheep and cattle grazed native grassland, which was highly modified in terms of function, structure and composition. In 1970, with the encroaching Canberra suburbs, a secure, people proof fence was erected around the site, enclosing a small population of kangaroos. Within the fenced area, sheep were set stocked as a means of controlling the pasture growth and reducing the risk of wildfire. The Navy also discretely kept the population of kangaroos in check until the encroachment of the surrounding suburbs prevented such management interventions. The removal of sheep resulted in an increase in the population of kangaroos, which ultimately caused declines in function, structure and composition, and endangered the habitats and populations of four threatened species, Striped Legless Lizard (*Dema impar*), Grassland Earless Dragon (*Tympanocryptis pinguicolla*), Golden Sun Moth (*Synemon plana*), and the Perunga Grasshopper (*Perunga ochracea*) present at the site (Hodgkinson 2014).

Despite the area being managed with the intent of enhancing or improving indicators of vegetation function, structure and composition, a population of kangaroos within the secure fenced area, rapidly increased between 2007–2010. Effects of over-grazing by kangaroos within this area became a sensitive environmental, social and political issue. Despite local campaigns to prevent the removal of the kangaroos, the kangaroo population was extirpated in 2011. This assessment (Fig. 15.4), shows there was no positive increase in vegetation condition in the period between 2011 and 2015, this can be attributed to the senior author having to rely on remote sensing (google earth) over this period to ascertain changes in the indicators, because access to the site was not permitted.

Transformation of the site of the former Belconnen Naval Transmitter Station is shown in Fig. 15.4. Between the 1830–60s, and before the landscape was subdivided as freehold land, shepherds grazed sheep and cattle without the aid of fences. Set stocking and continuous grazing commenced in the 1860s once fencing wire was introduced and the area was heavily grazed between 1900–1939. Early in this period the patches of woodland on higher ground were ringbarked to promote grass for grazing. Since the removal of those trees, any trees and shrubs were prevented from reestablishing and regenerating because of the heavy grazing pressure.

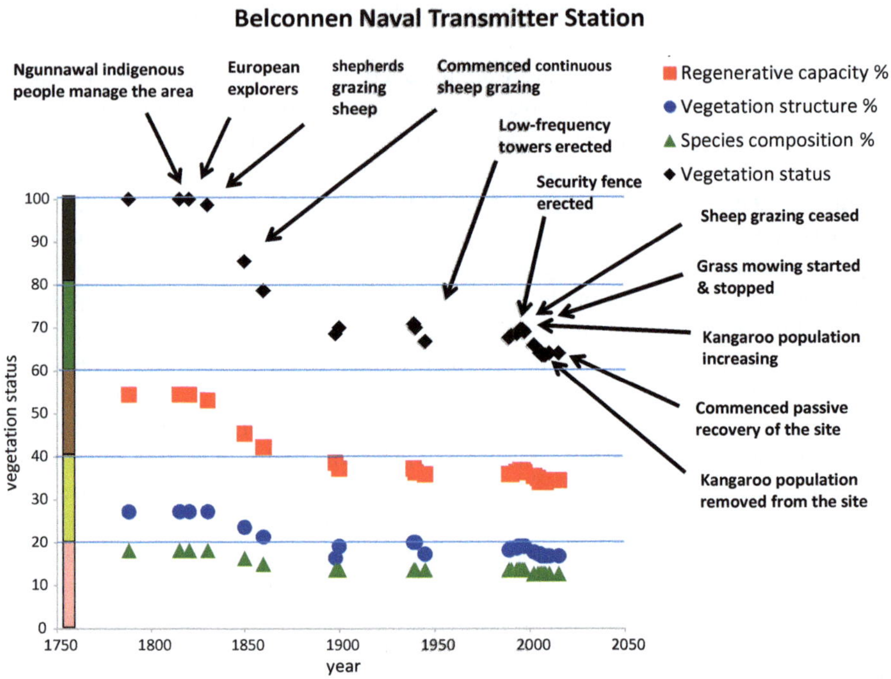

Fig. 15.4 Belconnen Naval Transmitter Station temperate grassland site. Showing land management practices effects on the structure, composition and function/regenerative capacity of the site

Set stocking or continuous grazing during that pastoral grazing period, 1860s-1939, resulted in pastures that were in low height, had low ground cover and were reduced in structural complexity. That grazing management regime also resulted in considerable bare ground during dry periods and extended droughts. The physical properties of the soil surface were changed with reduced infiltration and reduced available water to plants. Periods of strong winds and intensive rainstorms resulted in the erosion of topsoil and the loss of humus and initiated gully erosion. Preferential grazing, over those many decades, removed much of the floristic and functional plant diversity leading to the loss of many species of herbs and palatable grasses.

Since removal of the kangaroos in 2011, and in the absence of ground cover management such as autumn controlled burning, dense grass swards continue to suppress the survival of herbs. Anecdotal evidence indicates that the populations of the four threatened wildlife species have as yet not been affected by this increase in ground cover (Hodgkinson pers. comm).

This assessment (Fig. 15.4), shows the pastoral grazing period, 1860s-1939, had a relatively greater impact on the ecosystem components of function, structure and composition, than the overgrazing by kangaroos, 2007–2010.

3.3 Royal Australian Air Force Base Amberley

The Royal Australian Air Force (RAAF) Base Amberley is located west of Ipswich, Queensland and covers approximately 2500 ha. It became operational in 1940 when it received its first bomber aircraft and is one of the Air Force's longest-serving and most intensely used part of the Defence estate. It is currently being developed with massive infrastructure investments.

While the footprint of the Base has remained largely unchanged, additional areas of land have been acquired as buffers. The *area of interest* in the case study (also called Hansen's Farm or Hansen's Conservation Area by Defence until recently) was acquired in the early 1990s and lies to the north west of the runway and south of the Bremer River.

In 2016 Defence requested a demonstration of the VAST-2 approach to inform management on the resilience of the native vegetation in the north west of the Base in an area defined as an *area of interest* (Thackway and Pearson 2017). The *area of interest* comprises about 200 ha of the Base and has value as a buffer place for compensatory plantings and potential offset areas to exchange for areas required for new Defence infrastructure. Environmental offsets under the EPBC Act are reliant on reliable and high quality evidence-based policy and management decisions (DEE 2012). VAST-2 can provide such evidence.

The *area of interest* is currently managed for the following purposes:

a) as a buffer between the Base, the floodplain and encroaching urbanisation, pending major investments in Ipswich By-Pass;
b) as a biodiversity offset for major infrastructure as the Base is transformed from facilities largely built in the post WW2 period into a super base for a wider range of functions including fast jets (F/A-18F Super Hornet, EA-18G Growlers) and large jets (refuellers, heavy and battlefield airlift); and
c) for low intensity, limited military training (small infantry groups on foot).

The reference state for vegetation in south-eastern Australia can with confidence be seen as a legacy of indigenous local management that demonstrated flair for fire-stick farming at a landscape scale over very long periods (Gammage 2011). An intense European style agricultural land use locally and across the region has variously modified and replaced the original plant community at this site. The site is described as *Eucalyptus tereticornis* woodland on Quaternary alluvium (Regional Ecosystem type: 12.3.3 in local parlance), however the natural or pre-European reference state was not found on the present site or elsewhere on the Base. So, the reference state was a synthetic one, constructed from information collected from the remnant patches of *E. teretecornis* plant community found on the Base, and in the literature.

Land use and management decision-making for this area of interest was traced back to the large pastoral stations of the 1840s (Thackway and Pearson 2017). The Queensland Government's initiative in the 1850s to stimulate closer settlement resulted in the breakup up of large pastoral stations into blocks. Today there remains a strong spatial pattern between the density of the woody overstorey and the pattern of the boundaries of the 30–50-acre blocks delineated in the 1885 map of the area. It could not be established whether the small holdings of 30 to 50 acres were amalgamated to larger farms prior to Department of Defence acquiring the land in 1990–92.

The foci for this investigation are the current treeless areas within the area of interest. Historically, these areas were cleared and intensively managed for many years for agricultural production; initially cropping and more recently cattle grazing. The treeless areas in the landscape, in some cases, correspond to land parcel boundaries shown in the 1850s cadastral boundaries map. Air photographs and brief reconnaissance field survey showed that despite many of the fences having been removed after 1990, fence line boundary differences in vegetation continue to be observable. Despite the removal of domestic stock there was little or no evidence of recruitment of native shrubs and trees. Compared to a nearby remnant patch of *E. teretecornis*, these former agricultural soils were relatively hard and lacked evidence of biological activity suggesting reduced ecological function.

The absence of accessible records, prior to 1990–92 or for the area of interest as a whole after 1992, has resulted in major gaps and discontinuities in the environmental history that might be filled in future, and highlights the need to gather evidence and archive it as land managers change. Faced with the lack of records and direct measures of condition (using field-based attributes), estimates of attributes were derived from expert elicitation, environmental histories, interviews with skilled subject specialists and relevant metrics derived from multi-spatial and multi-temporal remote sensing datasets. The treeless area adjacent to site 3, within the area of interest has been passively managed as part of RAAF Base Amberley. Management practices in this treeless area include surveying and regularly treating noxious weeds, hazard reduction burns and some opportunistic tree planting, but overall the treeless area has remained ecologically transformed and unchanged since 1990.

Figure 15.5 shows the 1850s condition of the site was unmodified and then ringbarking and clear-felling using axes converted the area to arable crop land (Modified – VAST class IV). The use of ploughs, such as a mouldboard plough, to prepare the soil for farming removed the reproductive potential of the *E. teretecornis* plant community and transformed the site's function, structure and composition. The effect of decades of ploughing and cropping has fundamentally impaired the soil hydrology, soil physical structure, the biological status by dramatically reducing the soil organic matter and the invertebrate soil recyclers and bioturbators. Regenerative capacity (function) did not drop to zero, unlike vegetation structure and composition because several of the function criteria remain, although they are impaired: e.g. soil hydrology, soil physical properties and soil biology. Vegetation structure and composition were removed from the site when the site was converted to intensive cropping.

Cropping ceased in the 1960s–70s and the area was grazed by domestic livestock on exotic and native pasture species. Removing the stock has had little impact on the recovery of the function, structure and composition of the locally indigenous

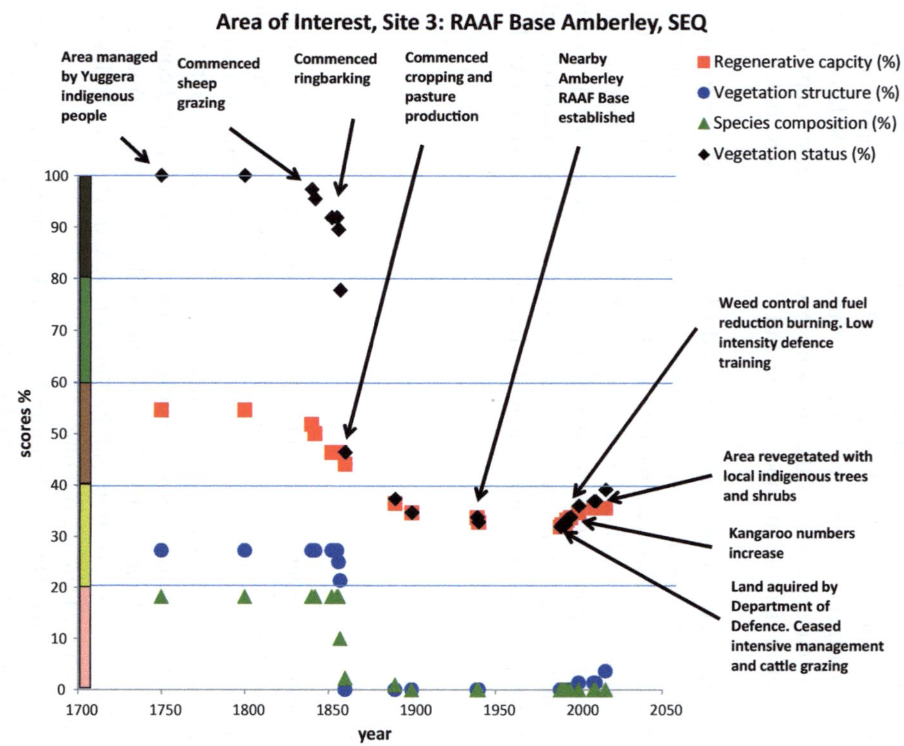

Fig. 15.5 RAAF Base Amberley area of interest (formerly known as Hansen's Farm) *Eucalyptus teretecornis* woodland on Quaternary alluvium. Showing land management practices effects on the structure, composition and function/regenerative capacity of the site. Comparing land management regimes between Defence facilities

understorey. It has however inadvertently increased total grazing pressure from kangaroos. Low intensity defence training activities appear to have little influence on the passive restoration of the structure, composition and function/regenerative capacity of the plant community.

Since the 1990s, Fig. 15.5 shows there is an observed increase in structure but not in species composition. This is because the height, cover and biomass of the ground layer increased following destocking but, because of the intensive cropping and grazing history, and with no augmentation of the area with native grass and forb species, species composition of the understorey has remained largely unchanged compared to the reference state. In addition, in 2009 the site was ripped and hand-sown with locally indigenous overstorey tree and shrub tube stock. Relative to the reference state, this enhancement of the site with overstorey species, represented only a small subset of the total number of tree and shrub species recorded in the reference state.

In 2009, a small part of the site was revegetated with locally indigenous species of trees and shrubs planted as tube stock. This site is in transition to a Transformed – VAST class III). The site illustrates the effects of major and enduring changes to structure, composition and function because of prolonged intensive cropping management. Intense development pressure on and off the RAAF Base is likely to drive up and generate new values for the area and there are opportunities to better describe and account for the effects of management on the condition of the RAAF Base Amberley.

4 Results

A comparison between the three case study sites and their respective land management regimes over time is shown in Table 15.3. All three sites had different land management histories and all sites were used for other land uses prior to each site being acquired for Defence purposes. All three sites are currently managed for enhancement or improvement of native vegetation function, structure and composition: Two sites, Tianjara Defence Training Area and Belconnen Naval Transmitter Station for biodiversity conservation and the third site RAAF Base Amberley continues to be managed for Defence training.

These two sites, Tianjara Defence Training Area and Belconnen Naval Transmitter Station, have also since first settlement by Europeans, retained their native vegetation. While the third site, RAAF Base Amberley was converted to non-native vegetation in 1860 and managed until 1990, initially for crop production and later for pasture production (land management regime #4) (Fig. 15.5).

During those periods when Tianjara Defence Training Area and Belconnen Naval Transmitter Station were managed for Defence purposes, range managers at both facilities reduced the biomass of the native vegetation using controlled fire (Tianjara) and sheep grazing and mowing (Belconnen) (land management regime #2), (Figs. 15.3 and 15.4).

Table 15.3 Periods of land management regimes in the three case study sites over time affecting native vegetation. Defence management is denoted as shaded areas

Land management regimes affecting native vegetation:		Tianjara defence training area	Belconnen naval transmitter station	Royal Australian Air Force Base Amberley
1	Practices that do not affect indicators of vegetation function, structure and composition	Non-Defence: 1788–1895	Non-Defence: 1788–1830	Non-Defence: 1788–1840
2	Practices that harvest/reduce vegetation products (biomass, fibre, flowers, fruit and nuts), which affect indicators of vegetation function, structure and composition	Non-Defence: 1896–1947 Defence: 1942–1983	Non-Defence: 1831–1939 Defence: 1940–2010	Non-Defence: 1841–1852 Non-Defence: 1939–1989
3	Practices that enhance or improve indicators of vegetation function, structure and composition	Non-Defence: 2004–2016	Non-Defence: 2011–2015	Defence: 1990-present
4	Practices that extirpate or remove indicators of the function, structure and composition	NA	NA	Non-Defence: 1853–1939
5	Practices that reconstruct or reinstate indicators of the function, structure and composition	NA	NA	Defence: 2009-present

Defence ceased managing the Tianjara Defence Training Area in the early 1980s and the area was transferred to the NSW National Parks Service. Commencing in the early 2000s, the area was managed to prevent frequent intense wildfire burning the area with the intent of enhancing or improving wildlife habitat including vegetation function, structure and composition (land management regime #3). The positive effects of controlling wildfire on vegetation condition, commencing in 2009, are shown in Fig. 15.3.

Defence ceased managing the Belconnen Naval Transmitter Station for military purposes and the site was decommissioned as a Defence facility in 2005. It should be noted that the previous pastoral grazing period, 1830s–1939, (i.e. land management regime #2) had a relatively greater impact on the ecosystem components of function, structure and composition, than the overgrazing by kangaroos, 2007–2010 (Fig. 15.4).

In comparison to the other two case studies, the RAAF Base Amberley site has had a more intensive and diverse history of land management regimes. Since 1990 Defence has managed the area with the intent of enhancing or improving indicators of vegetation function, structure and composition (land management regimes #3 and #5) however the site has not responded as rapidly as expected, despite the site being revegetated (Thackway and Pearson 2017). The slow recovery of the site (Fig. 15.5), can be attributed to historic intensive uses of the site such asland management regime #4 - cropping and cattle production. These historic management regimes have removed the capacity of the site to naturally recover and reinstate the original plant community, *Eucalyptus tereticornis* woodland. Decades of intensive land use and management regimes before Defence acquired the site, have dramatically affected the site's ecological function as shown by the very slow changes in the response indicators of the function, structure and composition (Fig. 15.5). Assisted restoration, by simply planting trees seedlings in 2009 and subsequently spot spraying weeds, has not been sufficient to 'trigger' improvements in the key indicators. To transform the site will require reconstructing the ecological function of the site, particularly the soil functional indicators.

5 Discussion

The results from three sites in the Australian Defence Estate highlight the value of applying an established consistent assessment framework that is independent and credible. The geographic frame is site specific, integrative and takes a past-present-future view that is easy for people to use in partnership with local stakeholders and ecologists. Site-based assessment scores are directly related to mapping modification classes at a landscape level. The development of comprehensive site histories is a precursor to assessing the transformation of the resilience of plant communities found at sites. They are useful engagement activities and the results suggest the value of rigorously evaluating the effects of Defence land management on the function, structure and composition of sites. It is no longer acceptable to assume that the current condition of a site reflects adequate management; new regulatory regimens, new geographic pressures, new values and rising expectations are clearly putting pressure on Defence estate management. Planning for future states is expected. Monitoring, evaluating and reporting the outcomes of land management practices is important for Defence's standing in the broader community, and would be critical to efficiently outsourcing environmental management of the Defence estate.

All three Defence facilities had a history of European land management that predated their acquisition and management for military purposes facilities. Historically, all three sites were agricultural, i.e. grazed with sheep and/or cattle to produce wool and/or meat. The area with the longest period of post-settlement land use and management history was Belconnen Naval Transmitter Station, ACT (187 years), followed by RAAF Base Amberley, Qld (177 years) and then by Tianjara Defence Training Area, NSW (158 years).

The three site histories of management represent the dynamics and drivers of the vegetation condition found across Defence facilities. These dynamics and drivers (shown on Figs. 15.3–15.5) show that Defence land management regimes have maintained or improved the condition of these former and current Defence estates. Because the information used in these assessments includes combined quantitative direct measurements and qualitative observations, the results underpinning these three figures do not lend themselves to statistical tests of significance. It should be noted that an issue facing the potential application of statistical assessments to determine the significance of change and trend across most land use types and land management units in Australia, is that most land managers, including Defence, do not regularly or systematically collect quantitative site-based data over time using standardised reproducible methods.

At a broader landscape scale, numerous Defence estates remain as largely natural areas, amid surrounding landscapes that are continually being transformed by relatively more intensive land use and management regimes (Thackway and Ford 2017). Two areas—Belconnen Naval Transmitter Station and Tianjara Defence Training Area—operated relatively briefly as Defence facilities, 65 years and 41 years respectively. Both sites were subsequently transferred to the relevant state or territory government to be managed as protected areas, in recognition that their prior maintenance and protection sustained important values including their biodiversity. In contrast, the third area, RAAF Base Amberley, started operating in 1990 and continues to this day with the base as a whole witnessing strong investment in new infrastructure to accommodate Defence's changing requirements and priorities. The treeless area, defined by Defence as an *area of interest*, has received limited investment in restoration, except for tree planting in 2009. Its restoration and rehabilitation are hampered by the previous and prolonged intensive cropping and pasture development, which has obviously affected the function, structure and composition of the site. In addition to rising amenity values and social expectations on the Defence estate, the area of interest is adjacent to new developments and there is prospect of compensatory or environmental offsetting investments that will be better informed by the assessment.

Together these three case studies highlight the value of geographically explicit approaches to the human and biophysical environments of the Defence estate. Using publicly available and credible methods to compile and assess specific representative land units and environmental management histories is a powerful tool for decision makers and managers. VAST-2 provides the social and scientific framework for gathering information and turning it into actionable knowledge of the effect of management (and future management) on the vegetation in the Defence estate. Knowing the impact of management on the structure, composition and function change relative to the reference state is a fundamental step in providing evidence that is meaningful to adaptive management, and compliance with regulatory and contractual arrangements.

Graphical summaries were derived for each case study site showing drivers of change and trends relative to a defined reference state. Systematic and comprehensive tabular chronologies of the land management history and the observed responses of the relevant response criteria and indicators are published for each site (e.g. Thackway 2012). Descriptive reports (e.g. Thackway and Pearson 2017) associated with each site provide an ecological interpretation enabling decision makers to quickly understand and assimilate complex ecological processes and their effects on landscape degradation, restoration and regeneration. The results assist decision-makers articulate the program logic for their management to achieve their outcomes, including maintenance, replacement, removal and recovery at site and landscape levels. The findings provide sound spatial and temporal insights into reconciling complimentary or competing land uses within the estate and for repurposing or restoring areas for new purposes.

It should be noted that the sites presented in this chapter present an historical account of the responses of criteria and indicators to management regimes (Table 15.2) and that no attempt has been made to provide recommendations for future management. However, should a land manager aim to enhance the function, structure and composition of a site, relative to a reference state, they might use a graph like those presented above, and consult an ecologist with the aim of making improvements in the condition of the site. For example, in regard to RAAF Base Amberley they might reasonably ask, "What would the likely outcomes be if a restorative land management regime was to be initiated?" Such a consultant might recommend that the land manager use controlled burns followed by tree planting and the introduction of inoculum. To track changes and trends and to continue reporting using the graphs, the ecologist might also recommend the site is monitored using fixed survey points and/or transects and a standardized set of criteria and indicators. Results from monitoring might be added to the assessment framework and assessed using the same criteria and indicators that were used to establish the historical graphical assessment (e.g. Figure 15.5).

Thackway and Freudenberger (2016) make an important distinction between the reference state and a contemporary baseline for a plant community under spatio-temporal land management regimes. They know that most environmental monitoring and tracking of the responses of plant communities to management seek to measure and observe change relative to the current baseline. In contrast, the VAST-2 framework reliably synthesises attribute information and well as measured observations to enable a direct comparison to the fully-natural reference state using indicators and criteria listed in Table 15.1.

Attempts to over-simplify the experience embodied in the VAST-2 assessment framework by trying to take shortcuts will be an opportunity wasted. Anticipated ways of shortcutting the framework and methodology, that may seem practicable, include: reducing the number of hierarchical levels to just criteria (Table 15.1); assessing fewer criteria (such as relying on just overstorey vegetation structure without reference to the underpinning indicators Table 15.1); skipping the systematic synthesis of land management practices and their effects on the function, structure and species composition; and only considering change relative to a baseline rather than a reference state. Given the value of reliable and credible knowledge that links management to environmental condition in the domains of social license, regulation and contracts, the anticipated shortcuts would pose a serious risk to the useability and reproducibility (and therefore comparative value) of any results. Properly applied the VAST-2 assessment framework is an open source, transparent, widely-tested, demonstrably workable and adaptable framework that has been successfully applied to improve Defence estate management. An advantage of the framework is that it requires a detailed and systemic chronology of land management practices to be compiled for a site along with relevant observations and direct measures of the effects of these management regimes practices on response indicators. These and other assessments have shown that the framework does not require in-depth specialist expertise, is not field and laboratory intensive, although where these data and information are available they can be readily incorporated into an assessment.

6 Conclusions

Systematic site-level assessments of change and trend in the condition of native vegetation on Defence land can provide valuable insights into how landscapes have been altered and might be deliberately altered in the future. This longer-term perspective, assessed relative to a fully natural reference state can then be used to help identify, and plan for, land management interventions to meet current and future strategic needs. The importance of community knowledge, values and expectations are better recognized. It provides opportunities for more credible, contestable evidence that can contribute to better management. This kind of rapid reliable and reproducible analysis can help answer such questions as: What is the condition of the native vegetation at different sites within a Defence facility relative to an accepted national standard? Can a regulator, Defence planner or land manager, for example, know with certainty how, why and when the condition of ecosystems of sites and landscapes was changed because of land management interventions? How can this knowledge be used by land managers to improve the condition of a site or landscape for its planned purposes? To consistently answer these questions across the wide range environments, a robust, transparent and iterative assessment framework was applied as an example to two former and one current sites in the Defence estate.

These assessments highlight unforeseen consequences of historical land management decisions. A lesson from such assessments is that they demonstrate the value of documenting such decisions so that those decisions and their effects on response indicators can be shared with the next generation of managers. These case studies provide a simple representation of the key historic drivers of land management practices on the structure, composition and function/regenerative capacity provides information in a rapid visual for everyone. Such results enable policy makers, planners and land managers to readily recognize and understand how complex socio-ecological drivers are affecting ecosystem services and dynamics. Such graphically-presented information should also be useful for informing planning decisions, strategic and tactical monitoring, analyzing and reporting the effects of Defence-related land management practices at the selected sites, as well as, for preparing compliance reports and environmental accounts of native vegetation condition (Cosier et al. 2015).

Acknowledgements The assistance of Robert Snedden is gratefully acknowledged through his extensive interviews of former users of the former Tianjara Defence Training Area. Robert also provided access to a digital copy of a rectified 1946 aerial photo of the impact area and a promotional Defence video "Tianjara Gunner", made in 1968, which is held at the Australian War Memorial. This information provided valuable insights into the effects that field firing had on the dry mallee-heath.

The support of Sue McIntyre and Ken Hodgkinson is appreciated, particularly their first hand on-ground field survey knowledge of the plant community found within the secure fence of the former Belconnen Naval Transmitter Station.

The assistance of Claire Joseph is gratefully acknowledged regarding RAAF Base Amberley through liaising with key Defence staff to enable access to reference material held on the base and escort during a brief reconnaissance of the area. Thanks to Frederick Ford for making available digitised, rectified time series aerial photos of RAAF Base Amberley.

References

Cooper, M. (2009). *Report on ACT lowland native grassland investigation*. Office of the Commissioner for Sustainability and the Environment, Canberra, Australia.

Cosier P., Davis, R., Grundy, M., Harding, R., Hillman, T., Possingham, H., Saunders, D., Sbrocchi, C., Smith, T., Thackway, R. & Thom, B., (2015). Evaluation of the Australian Regional Environmental Accounts Trial. Wentworth Group of Concerned Scientists, Sydney, Australia.

DEE. (2012). EPBC Act environmental offsets policy. Department of Environment and Energy, formerly Department of Sustainability, Environment, Water, Population and Communities, Canberra, Australia.

DoD. (2016). Defence environmental policy, strategy and plan. Department of Defence. Online at: http://www.defence.gov.au/estatemanagement/governance/Policy/Environment/Default.asp. [Accessed 6 Jun 2017].

DoD. (2017). Environmental factor management – Garrison and Estate Management System (GEMS). Department of Defence, Canberra, Australia.

Gammage, W. (2011). *The biggest estate on earth: how Aborigines made Australia*. Crows Nest: Allen & Unwin.

Gunn, R. H. (1983). *Lands of the Tianjara Army Training Area N.S.W.* Canberra: CSIRO.

Hilder, W. (1982). Historical background. In *Pigeon house and beyond – A guide to the Budawang Range and environs* (pp. 19–42). Silverwater: Produced by the Budwang Committee. Rodenprint.

Hodgkinson, K. C. (2014). *Condition of selected natural temperate grassland sites in urban and peri-urban Canberra*. Commissioner for sustainability and the environment ACT, Canberra, Australia.

Lesslie, R., Thackway, R., & Smith, J. (2010). *A national-level Vegetation Assets, States and Transitions (VAST) dataset for Australia (version 2.0)*. Bureau of Rural Sciences, Canberra, Australia.

Noss, R. F. (1990). Indicators for monitoring biodiversity: A hierarchical approach. *Conservation Biology, 4*, 355–364.

NPWS. (2002). *Wollongong LGA Bioregional Assessment (Part I): Native Vegetation of the Illawarra Escarpment and Coastal Plain. Report for the Commission of Inquiry into Planning and Development on the Illawarra Escarpment*. Sydney: NSW National Parks and Wildlife Service.

Sbrocchi C., Davis R., Grundy M., Harding R., Hillman T., Mount R., Possingham H., Saunders D., Smith T., Thackway R., Thom B., and Cosier, P. (2015) *Evaluation of the Australian Regional Environmental Accounts Trial*. Wentworth Group of Concerned Scientists, Sydney. ISBN 978-0-9944577-0-7.

Sneddon, R. C. (2016). *Tianjara: Early exploration, settlement and military heritage at Tianjara*. Canberra: RC Sneddon.

Thackway, R. (1985). *Analysis of techniques for mapping environments for fauna survey*. Canberra: Australian National University.

Thackway, R. (2012). *Transformation of Australia's vegetated landscapes, Belconnen Naval Transmitter Station*. Canberra, Australia. https://doi.org/10.4227/05/5088F7E9B9CCE.

Thackway, R. (2014). *VAST-2 Tracking the transformation of vegetated landscapes, Handbook for recording site-based effects of land use and land management practices on the condition of native plant communities, Version 3.0* The University of Queensland, Brisbane, Australia.

Thackway, R. (2016). Tracking Anthropogenic influences on the condition of plant communities at sites and landscape scales. In A. Almusaed (Ed.), *Landscape ecology-The influences of land use and anthropogenic impacts of landscape creation*. London: InTech. https://doi.org/10.5772/62874.

Thackway, R., & Ford, F. (2017). Characterising the Environmenal Values of the National Defence Estate, with emphasis on Native Vegetation. In S. Pearson, J. Holloway, & R. Thackway (Eds.), *Australian Contributions to Strategic and Military Geography*. New York: Springer.

Thackway, R., & Freudenberger, D. (2016). Accounting for the Drivers that Degrade and Restore Landscape Functions in Australia. *Land, 5*, 40. https://doi.org/10.3390/land5040040.

Thackway, R., & Lesslie, R. (2006). Reporting vegetation condition using the Vegetation Assets, States and Transitions (VAST) framework. *Ecological Management & Restoration, 7*, S53–S62.

Thackway, R., & Lesslie, R. (2008). Describing and mapping human-induced vegetation change in the Australian landscape. *Environmental Management, 42*, 572–590.

Thackway, R., & Pearson, S. (2017). *Assessing changes in the extent and condition of native vegetation on the Amberley RAAF Base – A pilot study*. Canberra: UNSW Canberra.

Thackway, R., & Specht, A. (2015). Synthesising the effects of land use on natural and managed landscapes. *Science of the Total Environment, 526*, 136–152.

Woodward, R. (2004). *Military geographies*. Oxford, UK: Blackwell.

Yapp, G., Walker, J., & Thackway, R. (2010). Linking vegetation type and condition to ecosystem goods and services. *Ecological Complexity, 7*, 292–301. https://doi.org/10.1016/j.ecocom.2010.04.008.

Index

S. Pearson et al. (eds.), *Australian Contributions to Strategic and Military Geography*, Advances in Military Geosciences,
https://doi.org/10.1007/978-3-319-73408-8

I

J

L

MIX
Papier aus verantwortungsvollen Quellen
Paper from responsible sources
FSC® C105338

If you have any concerns about our products,
you can contact us on
ProductSafety@springernature.com

In case Publisher is established outside the EU,
the EU authorized representative is:
Springer Nature Customer Service Center GmbH
Europaplatz 3, 69115 Heidelberg, Germany

Printed by Libri Plureos GmbH
in Hamburg, Germany